## First Edition

# Fire Investigator

Written by
Jon Jones

Edited by
Cynthia Brakhage
Carl Goodson

Validated by the International Fire Service Training Association
Published by Fire Protection Publications, Oklahoma State University

RECYCLABLE

*Cover: One of 3,300 "fire scenes" resulting from the Oakland, California, wildfire of October, 1991.
Photo by David M. Smith C.F.I., Associated Fire Consultants Inc., Tucson, Arizona.*

# The International Fire Service Training Association

The International Fire Service Training Association (IFSTA) was established in 1934 as a "nonprofit educational association of fire fighting personnel who are dedicated to upgrading fire fighting techniques and safety through training." To carry out the mission of IFSTA, Fire Protection Publications was established as an entity of Oklahoma State University. Fire Protection Publications' primary function is to publish and disseminate training texts as proposed and validated by IFSTA. As a secondary function, Fire Protection Publications researches, acquires, produces, and markets high-quality learning and teaching aids as consistent with IFSTA's mission.

The IFSTA Validation Conference is held the second full week in July. Committees of technical experts meet and work at the conference addressing the current standards of the National Fire Protection Association and other standard-making groups as applicable. The Validation Conference brings together individuals from several related and allied fields, such as:

- Key fire department executives and training officers
- Educators from colleges and universities
- Representatives from governmental agencies
- Delegates of firefighter associations and industrial organizations

Committee members are not paid nor are they reimbursed for their expenses by IFSTA or Fire Protection Publications. They participate because of commitment to the fire service and its future through training. Being on a committee is prestigious in the fire service community, and committee members are acknowledged leaders in their fields. This unique feature provides a close relationship between the International Fire Service Training Association and fire protection agencies which helps to correlate the efforts of all concerned.

IFSTA manuals are now the official teaching texts of most of the states and provinces of North America. Additionally, numerous U.S. and Canadian government agencies as well as other English-speaking countries have officially accepted the IFSTA manuals.

*ISBN 0-87939-182-0*        *Library of Congress LC# 00-106825*

*First Edition, First Printing, September 2000*        *Printed in the United States of America*

If you need additional information concerning the International Fire Service Training Association (IFSTA) or Fire Protection Publications, contact:

**Customer Service**, Fire Protection Publications, Oklahoma State University
930 North Willis, Stillwater, OK 74078-8045
800-654-4055    Fax: 405-744-8204

For assistance with training materials, to recommend material for inclusion in an IFSTA manual, or to ask questions or comment on manual content, contact:

**Editorial Department**, Fire Protection Publications, Oklahoma State University
930 North Willis, Stillwater, OK 74078-8045
405-744-5723    Fax: 405-744-8204    E-mail: editors@ifstafpp.okstate.edu

# Table of Contents

This first edition of IFSTA **Fire Investigator** provides fire investigators with the information, data, and resources necessary to meet the job performance requirements for fire investigators defined in NFPA 1033, *Standard for Professional Qualifications for Fire Investigator*. Fire investigators in every segment of the profession will find the information contained in this manual useful in their professional development. The ultimate goal of the manual is to promote the professional development of fire investigators who work in conjunction with fire departments around the world to reduce the loss of lives and property from fire.

Acknowledgment and special thanks are extended to the members of the IFSTA validation committee who contributed their time, wisdom, and talents to this manual.

## IFSTA Fire Investigator Validation Committee

**Chair**

Hugh Graham
Yates & Associates, Inc.
Atoka, OK

**Members**

James Arnold
Claremore Fire Department
Claremore, OK

Russell Chandler
Virginia Fire Marshal Academy
Richmond, VA

Terry Dawn Hewitt
McKenna Hewitt
Edmonton, Alberta

Andy Derrick
City of Merced Fire Department
Merced, CA

Charles Henrici
River Forest Fire Department
River Forest, IL

David Hooton
Spectrum Solutions
Signal Mountain, TN

Bill Peterson
Plano Fire Department
Plano, TX

Walter Robinson
NYS Office of Fire Prevention and Control
Hector, NY

G. Terry Smith
Illinois Fire Service Institute
Champaign, IL

Theresa McGehee
Durant, OK

Ron Rhoten
Stillwater, OK

David Smith
Associated Fire Consultants, Inc.
Tucson, AZ

Dennis W. Smith
Kodiak Enterprises, Inc.
Ft. Wayne, IN

**Contract Writer**

Jon Jones
Jon Jones & Associates
Lunenburg, MA

The following individuals and organizations contributed information or photographs or otherwise provided assistance that made the completion of this manual possible:

Donny Howard, Owasso, OK

Ed Prendergast, Chicago, IL

Daryl Cline, Tulsa (OK) Fire Department

Curtis Ozment, Tulsa (OK) Fire Department

Personnel of Engines 10-A and 16-A, Tulsa (OK) Fire Department

Joseph J. Marino, New Britain, CT

Ron Rhoten, Stillwater, OK

Steve George, OSU Fire Service Training

Bobby Henry, Altus (OK) Fire Department

G. Terry Smith, Champaign, IL

David Smith, Tucson, AZ

Elk Grove (IL) Village Fire Department

Russ Chandler, Richmond, VA

National Fire Protection Association, Quincy, MA

Shell Oil Co.

Bill Lellis, Larkspur (CA) Fire Department (Ret.)

Additionally, sincere gratitude is extended to the following members of the Fire Protection Publications staff whose contributions made the final publication of this manual possible:

**Project Manager/Editor**
Cynthia Brakhage, Associate Editor

**Technical Reviewer**
Carl Goodson, Senior Technical Editor

**Contract Writer**
Jon Jones

**Proofreaders**
Carol Smith, Senior Editor
Marsha Sneed, Associate Editor

**Research Technician**
Bob Crowe

**Editorial Assistant**
Tara Gladden

**Illustrators and Layout Designers**
Desa Porter, Senior Graphic Designer
Ann Moffat, Graphic Design Analyst
Ben Brock, Senior Graphic Designer

**Production Coordinator**
Don Davis, Coordinator, Publications Production

# Introduction

Fire investigation is the compilation and analysis of information related to fires and explosions. A fire investigation is usually conducted to determine and document the area of origin and cause of a fire and the factors that contributed to the ignition and subsequent growth. The information developed during the investigation is then used in many ways. In its basic form, it can be used to complete a fire incident report for the incident. The information, along with other data developed by a jurisdiction, can be used to identify trends, assist in the prevention of similar incidents, and develop public education programs. On a broader level, investigation data is used in the development of codes and standards aimed at reducing fire losses and preventing fire deaths.

Fire investigations are also conducted to determine whether a fire was intentionally set. In these cases, the cause and origin information developed by the fire investigator becomes an integral part of any criminal investigation and subsequent legal actions taken against the responsible parties.

 ## The Fire Investigator

As a result of the increase in the amount of information available on the ignition, growth, and development of fires, the amount of information a fire investigator must posses has dramatically increased in recent decades. Today's fire investigators must be knowledgeable in many areas including the science of fire, building construction and systems, hazardous materials recognition, evidence collection, scene documentation, and many other topics.

The field of fire investigation has changed in many ways. Historically, fire investigators were taught rules of thumb related to burn patterns and other physical evidence commonly found at the fire scene. For instance, it was a commonly accepted "rule" that the collapse of furniture springs was valid indication of exposure to a specific heat source. Many other rules of thumb were commonly taught to investigators. These concepts were readily accepted as fact. However, based on decades of research, we now know that the metal in furniture springs anneals by exposure to high temperatures for short times or lower temperatures for longer periods. The result is the same no matter how the heat was applied — the metal springs lose their springiness. To be successful, today's fire investigators can no longer apply these "rules of thumb" in their analysis but must rely on scientific principles and available research to reach their conclusions.

NFPA 1033, *Standard for Professional Qualifications for Fire Investigator*, defines the job of the fire investigator. That document defines the basic job performance requirements (JPRs) for fire investigators. It focuses on determining the cause and responsibility for fires. Sometimes, investigations may also assist in determining factors related to damage after a fire began (developing fire spread scenarios); identifying specific reasons for injuries or deaths (evaluating life safety issues); and assessing human fault or responsibility factors (determining facts related to potential negligence). These are often important functions; however, they are not the focus of this manual. If such functions are assigned, additional references and resources should be consulted.

**Purpose and Scope**

The purpose of this manual is to provide fire investigators with the information, data, and resources necessary to meet the job performance requirements for fire investigators as defined in NFPA 1033. Fire investigators in every segment of the profession will find the information contained in this manual useful in their professional development. Every effort has been made to base the material in the manual on the most current information available. The manual is intended for use as a text for formal training courses in fire investigation and as a guide for self-study by individual investigators. The ultimate goal of the manual is to promote the professional development of fire investigators who work in conjunction with fire departments around the world to reduce the loss of lives and property from fire.

**Notice on Gender Usage**

In order to keep sentences uncluttered and easy to read, this text has been written using the masculine gender, rather than both the masculine and female gender pronouns. Years ago, it was traditional to use masculine pronouns to refer to both sexes in a neutral way. This usage is applied to this manual for the purposes of brevity and is not intended to address only one gender.

# Chapter 1

# Organization, Responsibilities, and Authority

**Performance Objectives**

This chapter provides information that addresses performance objectives described in NFPA 1033, *Standard for Professional Qualifications for Fire Investigator*, particularly those referenced in the following sections:

**Chapter 3  Fire Investigator**
3-1.3
3-1.4

Organized fire departments have been reported in history since Augustus became the ruler of Rome in 24 B.C. and created the "vigil" service along with fire prevention regulations. Fire protection in North America has been around since the mid-1600s when communities began to assign fire suppression and prevention tasks to specific individuals. For many years, fire departments were primarily focused on the suppression of hostile fires in a community. The role of fire departments has gradually expanded throughout the years, and departments today provide a wide variety of services to their communities. One of these services is the investigation of fires in their jurisdiction. Fire investigators may be assigned to a separate unit, work in the fire prevention division, or work for a police agency.

In many states and provinces, the head of the fire department (fire chief) is given the responsibility of investigating the cause of fires as part of the legislation that empowers the department. As a result, many departments are legally obligated to conduct an investigation into the origin and cause of fires to which they respond. To meet these obligations, the department may use firefighters or fire officers who have been cross trained as investigators/fire inspectors to also perform fire investigations, or they may use members of their department who are specifically assigned as fire investigators. Again, based on the jurisdiction and the empowering legislation, fire investigators may also have police powers in the jurisdiction. In some areas, the investigation of fires may be the role of the county, state, or provincial fire marshal. In these cases, the local fire department may only play a support role in the investigation. This manual is written for those individuals who have the primary responsibility for the investigation of fires within a jurisdiction. Other companion manuals from IFSTA such as **Essentials of Fire Fighting** and **Introduction to Fire Origin and Cause** are directed at those individuals who support the fire investigator.

No matter what organizational structure the fire investigator works in, his primary function is to determine the cause of fires that occur within his jurisdiction. This data collected from investigations is used to develop fire prevention priorities, public education programs, and to plan for future fire protection needs in the community. The data is also combined with data from other jurisdictions to identify trends and to assist in the development and revision of fire-prevention-related codes and standards. Another major function of the fire investigator is the detection and investigation of incendiary fires. The thorough investigation of all fires that occur within a jurisdiction is a major deterrent to the crime of arson.

Many jurisdictions have organized investigative units that are made up of both investigators from the fire department and the police department. These joint investigative units can work very well as they foster a better working relationship between departments and provide access to the resources available to both agencies.

 ## Liaison with Other Agencies

The fire investigator regularly works with other agencies and organizations in the performance of his job (Figure 1.1). These liaisons may be very formal as in a joint agency investigative unit, or they may be informal as with a local construction contractor who provides heavy equipment to assist with debris removal on an as-needed basis.

The most common liaisons fire investigators need are those with agencies that are directly involved in their investigations. These agencies include the state/provincial fire marshal's office; local, state/provincial, and federal law enforcement agencies that have jurisdiction; the forensics laboratory that supports the jurisdiction; and the medical exam-

**Figure 1.1** Investigators often cooperate with other agencies. *Courtesy of Ron Jeffers.*

iner/coroner that serves the jurisdiction. In many jurisdictions, the fire marshal's office may need to be notified in the event of a serious fire/burn injury or death or a large property-loss fire. Should the incident under investigation involve a fatality, the medical examiner/coroner who serves the jurisdiction will be involved in the investigation. Incidents that may involve criminal activity result in the fire investigator working with the law enforcement agencies that have jurisdiction. Incidents with special circumstances, such as explosions or fires in government facilities or government-regulated facilities, may also necessitate that the fire investigator work in collaboration with state/provincial or federal law enforcement agencies that have jurisdiction.

During the course of an investigation, the fire investigator may often need information that is available from the municipal or state/provincial government. The fire investigator should have a good working relationship with the town or city clerk's office, the assessor's office, and the treasurer's office. These elements of the municipal government can assist with records and files that are essential to an investigation. Many times these municipal departments have counterparts at the state/provincial level of government. A working relationship with local officials can be helpful when additional information or resources are necessary at the higher governmental level.

Investigators may also work with the building department to obtain information regarding buildings involved in their investigations or for expertise on construction and potential hazards on the fireground. The local street or highway department may assist with background information related to streets within the jurisdiction and may also be able to assist in securing larger scenes with road barricades. The street or highway department may also be a source of heavy equipment to assist with debris removal during the examination of the scene.

The fire investigator and the agency he works for also maintain liaisons with private businesses that may be able to assist in the investigation of fires. One of the most important liaisons with the private sector is with the insurance industry. The insurance industry offers the following services:

- Provides information regarding the specific property on which the fire investigator may be working.

- Funds the collection of insurance-related data such as the Property Insurance Loss Register (PILR) that collects loss information as a result of fire, theft, and burglary. This database contains information regarding the company that paid a claim, the individual to whom the claim was paid, and the type of claim it was.

- Assists or collaborates with the public agency in a major fire investigation.

- Supplies specialized equipment, such as cranes and other heavy equipment, needed to remove debris from large buildings.

- Offers special experts or laboratory resources that may not be normally available to the fire investigator.

- Provides a source of educational materials related to arson and fire prevention (see Appendix A for listing).

The resources and assistance available from the insurance industry are essential for many public agencies in view of the financial situation in which they find themselves today.

Other private resources the fire investigator may want to maintain a liaison with include:

- Local financial institutions — information regarding mortgages and business history

- Utility companies—information regarding their distribution system and customer databases

- Local construction companies — sources of heavy equipment and expertise for debris removal and building stabilization

In many areas, arson task forces have been established to deal with intentionally set fires in the jurisdiction. The United States Fire Administration defines an *arson task force* as "... a management system for the purpose of developing and implementing strategies to control and prevent arson." Task forces can be legal or quasi-legal bodies or private advisory committees, but their purpose — fighting arson in a community—is the same. Arson task forces have the following functions:

- Mobilize public and private resources.

- Identify and coordinate responsibilities.

- Set policy or guidelines.

- Integrate the efforts of agencies, groups, and individuals who seek to prevent and control arson.

These organizations normally have policy-setting and program-implementation responsibilities and are not investigative units. For additional information on arson task forces, see *Arson Resource Directory*, a publication of the United States Fire Administration (FA-74/April 1993).

Fire investigators may also be faced from time to time with situations or investigations that are beyond their capacity to undertake. In these circumstances, jurisdictions may seek to establish an arson strike force. The United States Fire Administration defines an *arson strike force* as "... a special purpose, short-term mobilization of a team (or teams) of investigators together with allied resources that applies high intensity investigative efforts to a major arson incident or series of incidents." Thus, while arson task forces are policy-making bodies, an arson strike force is an operational investigative unit that has the ability to bring additional resources and capabilities into a major investigation. Arson strike forces not only help in the formal investigation but they assist investigators by sharing information about arsonists, their methods of operation, and special investigative techniques that may be used to apprehend and prosecute them. Additional information regarding arson task forces can be found in *Establishing an Arson Strike Force*, a publication of the United States Fire Administration (FA-88/ February 1989).

 ## Fire Investigator Professional Development

Fire investigation requires that the investigator have some knowledge of science, building construction, human behavior, forensics, law, and many other subjects. Fire investigators must also be able to write and speak clearly, perform strenuous physical labor, and organize and analyze large volumes of information. Today's fire investigator should also expect to have his or her findings and professional qualifications vigorously challenged by opposing counsel whenever cases go to court. Thus, to properly perform their job, fire investigators must be properly trained, constantly remain current with changes in their profession, and seek credentials that attest to their knowledge and abilities.

NFPA 1033, *Standard for Professional Qualifications for Fire Investigator*, establishes the minimum job performance requirements for the fire investigator. As with many of the other professional qualifications standards produced by the NFPA, this document is the basis for most fire investigator training programs in North America. The intent of this manual is to provide the fire investigator with information he needs to meet the requirements of NFPA 1033. Formal training programs are offered by many jurisdictions, state and provincial fire service training academies, national law enforcement academies, and the National Fire Academy. The International Association of Arson Investigators and other national associations of fire investigators also offer continuing education and formal certification programs based on NFPA 1033.

---

**Training and Certification for Fire Investigators**

The training and certification of fire service professionals has been centered around the NFPA Professional Qualifications Standards first issued in the early 1970s. As part of the Professional Qualifications System, NFPA 1033 defines the minimum job performance requirements for the fire investigator. The standard is periodically revised by the Association to meet changing fire investigation needs.

States, provinces, and other agencies such as the U.S. Department of Defense can adopt the professional qualifications standards and use them to develop training and certification programs for the fire service. Certification programs are designed to evaluate that an individual has the knowledge and skills necessary to perform a specific job. The professional qualifications standard for the specific job sets the criteria used for these evaluations — normally written tests and skill evaluations. As part of the professional qualifications system, organizations that provide certification to fire service professionals can seek to be accredited. Accreditation is the formal review by a third-party of the policies, procedures, and process used by an organization providing certification. There are currently two fire service accrediting organizations — the National Professional Qualifications Board and the International Fire Service Accreditation Congress. While there are only a few states and provinces currently accredited to certify fire investigators using NFPA 1033, the accreditation systems are in place and growing, and it is expected that more entities will seek accreditation to certify for this job. There are several states and provinces that offer fire investigator certification that is based on training and criteria established by the state/province.

Recently, the National Fire Academy issued several packages developed around the NFPA 1033 requirements for the fire service. These materials should help to increase the delivery of fire investigator training programs at the local level.

Two national organizations serving fire investigators, the National Association of Fire Investigators (NAFI) and the International Association of Arson Investigators (IAAI), also provide training and certification of investigators. Both national organizations and their local chapters provide training opportunities at various locations. These programs are normally focused on one or more aspects of the fire investigator's job and many times are directly related to the requirements of NFPA 1033 and the information covered in NFPA 921, *Guide for Fire and Explosion Investigations* (1995). Through its national certification board, NAFI administers the Certified Fire and Explosion Investigator program. This certification is based on an examination and is designed for individuals who are or will be actively involved in fire and explosion investigation. The IAAI administers the Certified Fire Investigator program. This program is aimed at identifying and recognizing an investigator's expertise. It is based on an evaluation of the individual's investigation training and related experience and a comprehensive examination. The IAAI program is nationally accredited to certify fire investigators by the National Professional Qualifications Board.

Additional information regarding training and certification can be obtained by contacting the following:

National Association of Fire Investigators
P.O. Box 957257
Hoffman Estates, IL 60195-7257

International Association of Arson Investigators
300 S. Broadway
Suite 100
St. Louis, MO 63102

---

Because of the rapid changes in this profession that are brought about by better technology and because of changes in codes and legal decisions, fire investigators constantly learn about their profession. To withstand the careful examination of their professional qualifications that takes place when they become involved with the legal system, fire investigators should also document their training and develop their professional credentials by seeking certifications in their field.

 ## Legal Foundations

Fire investigators must have the legal authority to conduct a fire investigation, and they must conduct themselves and their investigations in accordance with prescribed legal procedures. This is true for public-sector and private-sector investigators and for criminal and civil cases. It is also true whether the investigation is conducted in the United States or in Canada.

### Due Process

The first premise is that the constitution of each country defines *liberty* as the individual's freedom from governmental regulation. With few exceptions, constitutions do not govern rights between private persons.

There are two provisions in the American Constitution known as the "due process" clauses that protect individual rights and have serious implications in fire investigations. These clauses are found in the Fifth and Fourteenth Amendments. The Fifth Amendment provides, "No person shall ... be deprived of life, liberty or property without due process of law ..." The Fourteenth Amendment adds, "... nor shall any State deprive any person of life, liberty, or property without due process of law..." Due process requires that before a person's life, liberty, or property can be taken away, certain procedures that conform with recognized standards of fairness must be followed.

In Canada, the rights of individuals are protected by the *Canadian Charter of Rights and Freedoms*, which was enacted as part of Canada's constitution in 1982. Section 7 of the Charter is the counterpart of the due process clauses in the American Constitution. It provides, "Everyone has the right to life, liberty and security of the person and the right not to be deprived thereof except in accordance with the principles of fundamental justice." The words of Section 7 of the Charter are similar to the Fifth Amendment, though the development and application of the law arising from this section is somewhat different.

In either country, the result of the constitutional protection of individual rights is the regulation of certain powers exercised by fire investigators, such as entering a fire scene, collecting evidence, and interviewing or taking statements from witnesses or suspects. The fire investigator shall ensure that due process is served.

### Authority

In both countries, public-sector investigations must be conducted within the bounds of statutory authority. The constitution of each country defines the boundaries of a government's lawmaking power, balancing it with the constitutionally protected rights of the individual.

The investigation must be conducted within the boundaries of statutory authority. One or more statutes will apply to a given case and dictate the purpose of the investigation and the powers and jurisdiction of the investigator. If an investigator exceeds his or her authority, evidence from the investigation might be ruled inadmissible, charges might be dismissed, or in extreme cases, an investigator might be subject to criminal penalties or civil liability.

### Private-Sector Investigations

In the private sector, there are also laws governing legal procedures and the authority of investigators. Constitutional laws, such as due process clauses, have little to do with the private sector and civil litigation. The powers and legal authority of private-sector investigators are governed by civil laws governing rights and obligations between private persons. Rights or obligations arising from contracts, insurance policies, and tort laws such as negligence impact the authority of the private-sector investigator. State or provincial laws regulating the licensing of investigators also have an impact.

It is important that those conducting investigations for civil cases act within the bounds of their legal authority. Failure to do so can have various negative results. Evidence or statements might be ruled inadmissible in court, or an investigator might be exposed to civil suit or criminal prosecution. In the insurance arena, for example, bad faith cases are founded on circumstances where an investigator exceeded his authority.

While a detailed analysis of the legal foundations of private-sector fire investigations are beyond the scope of this book, some of the central legal issues that affect fire investigators in civil cases are introduced in later chapters of this manual.

##  Safety

While firefighter occupational health and safety has been the topic of considerable discussion and attention for some time, the health and safety of fire investigators has only recently begun to get attention. A significant part of the investigator's job is to protect his health and personal safety. While conducting investigations, fire investigators are regularly exposed to hazards that can result in immediate injury or, in some cases, long-term health problems.

NFPA 1500, *Standard on Fire Department Occupational Safety and Health Program*, requires that fire departments adopt a risk management plan that addresses all aspects of their operations. The standard defines the components of a risk management plan as follows:

• Risk identification

• Risk evaluation

• Risk control

• Risk management monitoring

One of the objectives of risk management is the reduction of risk to individuals from potentially harmful situations. From this list, it can be seen that this can be accomplished by identifying potential risks, evaluating the hazard once it is identified, and then working to control the risk.

The potential hazards (risks) that fire investigators face in the performance of their jobs include some that are specific to their job and many of those that face firefighters who are involved in suppression. The fact that many fire investigators work alone in potentially hazardous locations also increases their risk.

## Safety Hazards
The hazards of the job as well as precautions for the investigator are discussed in this section.

### Structural Hazards
Many buildings that fire investigators enter are unstable as a result of the fire damaging structural components, suppression activities that may have removed or weakened structural components, and the accumulation of water in the building. These conditions can lead to structural collapse or potential fall hazards for the investigator working in the building. Other hazards the investigator may encounter in the fire building are the accumulation of fire debris or rubble and reduced visibility due to smoke or lack of light. A basic knowledge of building construction is helpful when trying to assess the stability of a building (see Chapter 3, "The Basics of Building Construction as it Relates to the Investigator") (Figure 1.2).

### Hazardous Atmospheres
Products of combustion (smoke and hot gases) may still be present as the investigation begins. Other hazards include oxygen depletion in closed spaces and the accumulation of toxic gases, such as carbon monoxide, in the building. Depending on the occupancy of the fire building, there could also be materials in the atmosphere that the investigator might not anticipate such as pesticides in a barn or toxic gases in an industrial facility. These conditions can be identified using available monitoring equipment.

**Figure 1.2** Investigators must assess the structural integrity of a building before entering.

### Hazardous Materials

The fire investigator may face any number of hazardous materials in a fire building. These materials may be in the building for any number of reasons. They may be there as part of legitimate activities that normally take place in the building such as chemicals used in an industrial process. The more dangerous situations involve materials that are stored or used in locations in which they are not normally expected. The best examples of these situations are illegal drug production laboratories and homeowners who store dangerous materials in their homes or garages for unknown reasons. Materials, such as asbestos, that are normally encapsulated within the building may also be present and exposed within the fire building. The identification of potential hazardous materials is discussed in greater detail in Chapter 6, "The Investigator's Role In Recognizing Hazardous Materials."

### Standing Water

The water used for fire extinguishment and possibly from broken pipes in the fire building can result in several hazards to an investigator. As discussed earlier, it can add additional weight to the already weakened building causing a collapse. Water can also accumulate in low spots and result in dangerously deep pools that could entrap the investigator. Water accumulations can also increase the potential for slips and may increase the risk of electrical shock hazards if there are energized conductors nearby.

### Building Utilities

Utilities that make a building livable under normal conditions may result in hazards to fire investigators after a fire. Energized electrical conductors present shock hazards. Broken gas lines can result in explosive mixtures of gases in the building. Broken water pipes can add to the standing water hazard previously discussed.

All these hazards place the fire investigators at risk as they perform their job. This risk is increased by the vary nature of the job and how it is conducted. Many times fire investigators must enter fire buildings before complete extinguishment, thus exposing them to the hazards normally encountered by firefighters. In many cases, the fire investigator must work alone in buildings well after other fire department personnel leave the scene. In these situations, the investigator can be at risk of assault or be unable to contact assistance should he be injured while conducting the investigation.

While they are performing their job, fire investigators must always be alert for conditions or situations that could possibly place them at risk of injury or death. Knowing potential hazards and being able to identify them in the field are critical to the health and safety of all fire investigators.

## Reducing the Risk

Three common methods are available to the fire investigator to control or reduce the risks encountered on the job. These methods are avoidance, control, and transfer.

### Avoidance

This method is the most straightforward as it simply involves not taking the perceived risk. This could involve not going into an area of a building that is unstable or not working around exposed electrical conductors (Figure 1.3). While avoidance is often the most effective way to deal with a potential risk, it is often not practical. Fire investigators may need to perform tasks that involve some level of risk. In that case, they will use one of the other methods or a combination of the two to minimize their risk.

### Control

Risk control is the most common method used. Good risk control measures reduce the chance of an accident occurring and reduce the severity of those that do. Risk control methods include using

**Figure 1.3** Investigators must beware of dangerous conditions.

**Figure 1.4** Investigators control risk by working in teams. *Courtesy of Joseph J. Marino.*

proper personal protective clothing and equipment, working in teams, securing utilities before entering a building, providing proper lighting of the scene so that hazards can be seen, removing weakened portions of a structure, and shoring weakened structural members (walls, floors, ceilings, and roofs) (Figure 1.4).

### Transfer

This method of risk management is the transferring of risk to someone else. An example of risk transfer would occur on the scene of an active hazardous materials incident when the fire investigator uses members of a hazardous materials response team trained in the use of special protective equipment to survey a scene and secure potential evidence. The risk to the fire investigator not trained in the use of special protective equipment would be high but significantly lower for the specially trained hazardous materials response specialists. This method may not be the most desirable but has applications in special situations.

The safety of fire investigators is thus maintained through a combination of following the policies and procedures developed for the fire department or agency they represent, the ability to recognize potential hazards, and the fire investigators' ability to apply the proper risk management techniques to each fire scene on which they are called to work.

### ◆ Tools of the Profession

Fire investigators use a wide variety of tools and equipment in the performance of their job. The equipment ranges from very basic hand tools used to excavate the fire scene and collect samples to canine accelerant detection teams and at times very sophisticated computer modeling programs (these are discussed in detail later in this chapter). While the investigator may not personally use all the equipment that is available, there are some basic items that the investigator will use. The investigator must also understand what is available to assist with an investigation and when and how to access special services that are available.

### Personal Protective Equipment

The fire investigator uses a variety of personal protective clothing and equipment, depending on the conditions under which he is required to operate. At a very minimum, the investigator should be provided with and use safety shoes or boots, gloves, head protection, and protective clothing such as coveralls or structural fire fighting coat and pants. Depending on conditions, the fire investigator may also require respiratory protection such as a filter mask (rated for the use it is given) or self-contained breathing apparatus (SCBA) (Figure 1.5). Depending on the work being performed, eye protection may also be necessary (Figure 1.6). Special protective equipment, such as full encapsulation suits or line respirators used in incidents involving hazardous materials, should only be used by fire investigators who have been trained in its use.

### Standard Equipment and Tools

NFPA 1033 lists the following as the standard equipment and tool list for the fire investigator:

- Camera, with film and flash
- Flashlight
- Shovel
- Broom
- Hand tools
- Tape measure
- Evidence collection containers

All tools used by the fire investigator should be kept free of rust and cleaned after each use to prevent contamination. Proper cleaning methods are discussed later in this manual. Cleaning materials such as paper wipes or towels, liquid soap, and

**Figure 1.5** Investigators often need respiratory protection.

**Figure 1.6** Investigators should wear eye protection.

isopropyl alcohol should be available. The investigator should also have a supply of barrier tape and other materials for marking a fire scene and protecting potential evidence (Figure 1.7).

**Figure 1.7** The scene must be secured to protect evidence.

### The Fire Investigator's Tool Kit

As the fire investigator conducts a scene examination and then collects evidence, he often requires a set of hand tools to do the job properly. These tools should be used specifically for investigations, and to avoid contamination, they should be kept separate from those tools used by suppression crews for forcible entry and overhaul. The following tools are recommended as part of the fire investigator's personal tool kit (Figure 1.8):

- Claw hammer
- Hatchet
- Pry bar
- Hacksaw (extra blades)
- Keyhole saw (extra blades)
- Screwdrivers (multiple sizes and types)
- Pliers
- Wire cutters
- Utility knife (extra blades)
- Mason trowel
- Wood chisel
- Cold chisel

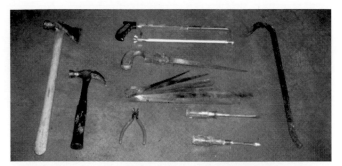

**Figure 1.8** A typical array of hand tools used by investigators.

- Brushes (paint or whisk broom)
- Tape measure
- 6" ruler
- Pencil magnet
- Tire depth-gauge tool
- Pencil scribe
- Marking pens
- Multi-meter (Volt/Ohm)
- A toolbox large enough to store and protect the tools

**Special Tools and Equipment**

NFPA 1033 defines special tools as: "Tools of a specialized or unique nature that may not be required for every fire investigation..."Examples of this type of equipment include heavy excavation equipment, ladders, rope, or portable lighting equipment. Depending on the scene, the investigator may require special equipment to safely and efficiently conduct the investigation. The fire investigator should know the policies of the organization regarding the use of specialized equipment and where it can be obtained when needed. In many cases, other agencies, such as the state/provincial fire marshals office or federal law enforcement agencies, can assist with special equipment for major incidents. In the case of major losses involving commercial, industrial, or similar properties, the insurance carrier or carriers may be willing to provide financial and other forms of support to accomplish the investigation.

Accelerant-detection canine teams and electronic hydrocarbon detectors would also fall into the category of special equipment. Well-trained canine teams have proven to be valuable investigative resources. The canine teams are most useful at scenes where the investigator suspects the use of an accelerant but the liquid cannot be smelled. Canine teams are particularly useful for locating potential sites to collect samples for laboratory analysis in heavily damaged or partially collapsed fire buildings.

 **Computer Fire Modeling**

Fire modeling programs have been available for many years, but with the advent of very powerful and affordable personal computers, these programs are more available to the fire investigator. The fire growth models available today are mathematical programs that can rapidly estimate changes in the environment caused by a fire in a room or space over a period of time. The changes in the environment may be in terms of smoke produced or temperature depending on the model used. Computer models can be used by trained individuals to assist in the reconstruction and analysis of fires by predicting the fire growth and smoke production and spread. The output of the models can assist the fire investigator in analyzing various fire growth scenarios based on the known outcome of the fire. Several major fire incidents in recent history, including the Dupont Plaza Hotel fire in San Juan and the Happy Land Social Club fire in New York City, have been reconstructed using computer models.

As with any analytical tool, computer fire models as investigative tools have limitations. The user must be very familiar with the model being used and be certain that the model is appropriate for the application, that the information (data) that is input is accurate, and that the output is properly interpreted. These tools can be useful in the reconstruction of incidents by providing the investigators with a method for testing their hypotheses regarding the ignition and development of the fire. Fire investigators should not attempt to use these tools unless they have been trained in their use, limitations, and the analysis of their output. However, because of their potential value and availability, the fire investigator should become familiar with the models and develop a basic understanding of the information they produce.

**Common Computer Fire Models**
**ASET-B**

This program, designed to operate on personal computers, calculates the temperature and position of the hot, upper smoke layer in a single room with closed doors and windows.

Input: Heat-loss, height of fire, room ceiling height, floor area of room, maximum time for simulation, and rate of heat release of the fire.

Output: Temperature and thickness of hot upper layer as a function of time.

### FPEtool

FPEtool is a collection of programs that are useful in estimating the potential fire hazard in buildings. The programs require a variety of different inputs depending on the specific portion of the program used.

Output: Temperatures, location, visual obscuration, oxygen content, and carbon monoxide and carbon dioxide concentrations in the smoke produced by the fire. Also provides estimated egress time, ignition of exposed items, smoke flow, fire, wind and stack pressures on doors, and activation times for detectors and sprinkler heads.

Because of its versatility and relative ease of use, FPEtool is very popular with the engineering and investigation communities.

### Hazard I

This program is designed to predict hazards to occupants of a building on fire. The program predicts the outcome of a fire in terms of the number of occupants who escape or are killed from a fire in a building.

Input: Building information, location of occupants, fire scenario data, and tenability criteria.

Output: The time of escape and the time, location, and probable cause of death for fatalities.

This program is very sensitive to the user input data and is judgmental and dependent on the expertise of the user. The use and analysis of the output require considerable expertise in the program and in fire protection. The program is complex and is not for the casual user.

### FASTLite

FASTLite is a new collection of programs that builds onto the core routines found in FPEtool and other modeling programs. The routines provided in the package include:

- Three-room fire model
- Egress time
- Heat and smoke detector activation
- Suppression by sprinklers
- Radiant ignition of nearby fuel

The program provides both graphic and spreadsheet output formats. It was developed to provide engineering calculations of fire phenomena including quantitative estimates of some likely consequences of a fire. As with all computer models, it is intended as only a supplement to the informed judgment of the qualified user.

 ## Conclusion

Being a professional fire investigator today is more challenging that at any other time. Fire investigators must have some understanding of fire behavior, physical science, building construction, human behavior, forensics, law, and many other subjects. They must strive to stay current on all the technological, scientific, and legal changes that occur each year. They must know the hazards associated with fire investigations and how to mitigate those hazards. Finally, they must know how to use the latest fire modeling software available to assist them in their investigations.

# Chapter 2

# Fire Behavior

## Performance Objectives

This chapter provides information that addresses performance objectives described in NFPA 1033, *Standard for Professional Qualifications for Fire Investigator*, particularly those referenced in the following sections:

**Chapter 3  Fire Investigator**
3-2.4 a
3-2.5 b

The investigation of fires requires the systematic analysis of information related to an incident. One of the key components of this analysis is an understanding of the science of fire and how it behaves from the time of ignition until it is extinguished. Fire investigators use this knowledge from the time they arrive on the scene and begin the initial examination until they complete their final report on the incident. Without a good understanding of fire behavior, the investigator cannot conduct a credible investigation.

Throughout history fire is a phenomenon that has been both a help and a hindrance to mankind. Fire has heated our homes, cooked our food, and helped us to become technologically advanced. Fire, in its hostile mode, has also been a danger to us for as long as we have used it. Technically, fire is a chemical reaction that requires fuel, oxygen, and heat to occur. Over the last 30 years, scientists and engineers have learned a vast amount about fire and its behavior.

This chapter introduces several basic concepts from physical science that impact the ignition and development of a fire. The information contained in this chapter can be used by the investigator to interpret what is observed at the fire. An understanding of fire behavior and the phases through which a growing fire passes assists the investigator in the analysis of what is observed on the scene.

##  Physical Science

*Fire* is a rapid chemical reaction that gives off energy and products of combustion that are very different in composition from the fuel and oxygen that combined to produce them. To understand the reaction we call fire, how it grows, and its products (products of combustion), we need to look at some basic concepts from physical science. *Physical science* is the study of the physical world around us and includes the sciences of chemistry and physics and the laws related to matter and energy. The basic science information in this section is referred to throughout this chapter as fire growth and development are discussed.

### Energy and Work

In the study of any science, energy is one of the most important concepts. *Energy* is simply defined as the capacity to perform work. Work occurs when a force is applied to an object over a distance. In

## Measurement Systems

In any discussion of science, information is presented using numbers. Investigators frequently use measurements while doing their jobs. They regularly use a measurement system to describe the size of a room or building (in feet or meters) or the volume of a container found at the fire scene (in gallons or liters). For these numbers to make any sense, they must be used with some unit of measurement that describes what is being measured. The units are based on a measurement system. In the United States the *English* or *Customary System* is commonly used. Most other nations and the scientific community use a form of the metric system called the *International System of Units* or *SI* (after the French Systeme International) as their measurement system.

Each system defines specific units of measure. Table 2.1 shows some of the base units used in each system.

A large variety of derived units are generated from these base units. For example, the base unit for length in SI is the meter (m). From this base unit, you can derive measurements for area in square meters ($m^2$) and volume in cubic meters ($m^3$). Measurements for speed can be derived from length and time and described in feet per second or meters per second (fps or m/s). In the discussion of fire behavior that follows, the derived units for heat, energy, work, and power are introduced and discussed. Other units used in the SI are hour (h) and liter (L). While they are not considered base units, they are widely accepted and used. While mass is considered a base unit, weight is not. Weight is the measurement of the gravitational attraction on a specific mass. In the Customary system, the unit for weight is the pound (lb). In the SI, weight is considered to be a force and is measured in newtons (N). Both Customary and SI units are provided throughout the chapter.

One of the reasons why the scientific community uses the SI is that it is a very logical and simple system based on powers of 10. This allows for the manipulation and conversion of units without the fractions we must use with the Customary System. For example, in the Customary System the unit of length is the foot. The other recognized units are the inch (1/12th of a foot), the yard (36 inches or 3 feet), and the mile (5,280 feet or 1,760 yards). In the SI, the unit of length is the meter. To express length in larger or smaller terms, the system uses the prefixes shown in Table 2.2. Thus, a centimeter is 1/100th of a meter, and a kilometer is 1,000 meters.

### Table 2.1
### The Base Units of Measurement

| Quantity | Customary System | SI System |
|---|---|---|
| Length | Foot (ft) | Meter (m) |
| Mass | | Kilogram (kg) |
| Time | Second (s) | Second (s) |
| Temperature | Fahrenheit (°F) | Celsius (°C) |
| Electric current | Ampere (A) | Ampere (A) |
| Amount of a substance | | Mole (mol) |
| Luminous intensity | | Candela (cd) |

### Table 2.2
### Names and Symbols for SI Prefixes

| Prefix | Symbol | Multiply By |
|---|---|---|
| Tera | T | $10^{12}$, 1 trillion |
| Giga | G | $10^9$, 1 billion |
| Mega | M | $10^6$, 1 million |
| Kilo | k | $10^3$, 1,000 |
| Deci | d | $10^{-1}$, one tenth |
| Centi | c | $10^{-2}$, one hundredth |
| Milli | m | $10^{-3}$, one thousandth |
| Micro | $\mu$ | $10^{-6}$, one millionth |
| Nano | n | $10^{-9}$, one billionth |
| Pico | p | $10^{-12}$, one trillionth |

other words, *work* is the transformation of energy from one form to another. The SI unit for work is the *joule* (J). The joule is a derived unit based on a force in expressed *newtons* (also a derived unit — kg m/$s^2$) and distance in *meters*. In the Customary System, the unit for work is the foot-pound (ft lb). Figure 2.1 shows an example of work being performed.

The many types of energy found in nature include the following:

- **Chemical** — Energy released as a result of a chemical reaction, such as combustion

**Figure 2.1** Moving debris is a form of work. *Courtesy of Joseph J. Marino.*

- **Mechanical** — Energy possessed by an object in motion, such as a rock rolling down a hill

- **Electrical** — Energy developed when electrons flow through a conductor.

- **Heat** — Energy transferred between two bodies of differing temperature, such as the sun and the earth

- **Light** — Visible radiation produced at the atomic level, such as a flame produced during the combustion reaction

- **Nuclear** — Energy released when atoms are split (fission) or joined together (fusion) (**NOTE:** Nuclear power plants generate power as a result of the fission of uranium 235.)

Energy exists in two states: kinetic and potential. *Kinetic energy* is the energy possessed by a moving object. *Potential energy* is the energy possessed by an object that can be released in the future. A rock on the edge of a cliff possesses potential mechanical energy. When the rock falls from the cliff, the potential energy is converted to kinetic energy. In a fire, fuel has potential chemical energy. As the fuel burns, the chemical energy is converted to kinetic energy in the form of heat and light.

## Power

*Power* is an amount of energy delivered over a given period of time. In our example for work, an investigator is shown moving fire debris over a distance during an investigation. The investigator was expending energy over a distance and thus performing work. If the time to complete the debris removal is known, then the amount of power required to perform the task could be determined. Throughout history, fire has been used to generate power in many ways. A fuel's potential energy is released during combustion and converted to kinetic energy to run a generator or turn a shaft that "powers" a machine. The derived units for power are horsepower (hp) in the Customary System and watts (W) in SI.

In the study of fire behavior, researchers frequently address power when they consider the rate at which various fuels or fuel packages (groups of fuels) release heat as they burn. During the last several decades, researchers at the National Institute of Standards and Technology have compiled a great deal of information on the heat release rates (HRRs) of many fuels and fuel packages. This information is very useful in the study of fire behavior because it provides data on just how much energy is released over time when various types of fuels are burned. Heat release rates for specific fuel packages are discussed in more detail in the Fire Development section later in the chapter.

## Heat and Temperature

Anyone who has ever fought a fire or even watched a fire fighting operation knows that there is a tremendous amount of heat generated. *Heat* is the energy transferred from one body to another when the temperatures of the bodies are different. Heat is the most common form of energy encountered on earth. *Temperature* is an indicator of heat and is the measure of the warmth or coldness of an object based on some standard. In most cases today, the standard is based on the freezing (32°F and 0°C) and boiling points (212°F and 100°C) of water. Temperature is measured using degrees Celsius (°C) in SI and degrees Fahrenheit (°F) in the Customary System (Figure 2.2).

The approved SI unit for all forms of energy including heat is the *joule*. While joules are used to describe heat in current literature, heat was described in terms of calories (Cal) or British thermal units (Btu) for many years. A *calorie* is the amount of heat required to raise the temperature of 1 gram of water 1 degree Celsius. The *British thermal unit* is the amount of heat required to raise the tempera-

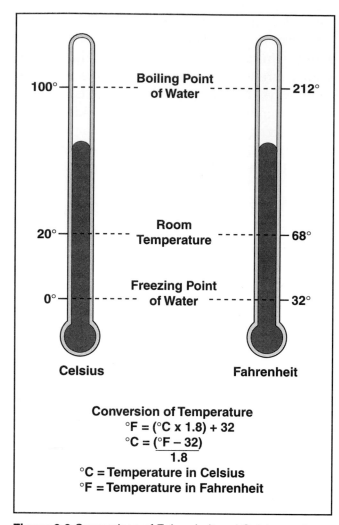

Figure 2.2 Comparison of Fahrenheit and Celsius scales.

ies must be at different temperatures. Heat moves from warmer objects to those that are cooler. The rate at which heat is transferred is related to the temperature differential of the bodies. The greater the temperature difference is between the bodies, the greater the transfer rate will be. The transfer of heat from body to body is measured as energy flow (heat) over time. In the SI, heat transfer is measured in kilowatts (kW). In the Customary System, the units are Btu per second (Btu/s). Both units (kW and Btu/s) are expressions that relate to power (see the discussion of mechanical equivalent of heat in the preceding Heat and Temperature section).

Heat can be transferred from one body to another by three mechanisms: *conduction, convection,* and *radiation.* Each of these is discussed in some detail in the following sections.

### Conduction

When one end of metal rod is heated with a flame, the heat travels throughout the rod. This transfer of energy is due to the increased activity of atoms within the object. As heat is applied to one end of the rod, atoms in that area begin to move faster than those in other areas. This activity causes an increase in the collisions of atoms. Each collision transfers energy to the atom being hit. The energy, in the form of heat, is transferred throughout the rod (Figure 2.3).

This type of heat transfer is called conduction. *Conduction* is the point-to-point transmission of

ture of 1 pound of water 1 degree Fahrenheit. The calorie and the Btu are not approved SI units but are still frequently used. The relationship between the calorie and the joule is called the *mechanical equivalent of heat,* where 1 calorie equals 4.187 joules and a Btu equals 1,055 joules.

## Transmission of Heat

The transfer of heat from one point or object to another is a basic concept in the study of fire. The transfer of heat from the initial fuel package to other fuels in and beyond the area of fire origin controls the growth of any fire. Investigators use their knowledge of heat transfer as they analyze the fire scene and determine the method of heat transfer from the fuels involved in the ignition to other fuels in the compartment or area of origin. The definition of heat makes it clear that for heat to be transferred from one body to another, the two bod-

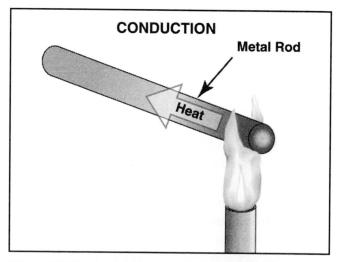

Figure 2.3 The temperature along the rod rises because of the increased movement of molecules from the heat of the flame.

heat energy. Conduction occurs when a body is heated as a result of direct contact with a heat source. Heat cannot be conducted through a vacuum because there is no medium for point-to-point contact.

In general, heat transfer early in the development of all fires is almost entirely due to conduction. Later, as the fire grows, hot gases begin to flow over objects some distance away from the point of ignition, and conduction again becomes a factor. The heat from the gases in direct contact with structural components or other fuel packages is transferred to the object by conduction.

Heat insulation is closely related to conduction. Insulating materials do their jobs primarily by slowing the conduction of heat between two bodies. Good insulators are materials that do not conduct heat well because of their physical makeup and thus disrupt the point-to-point transfer of heat energy. The best commercial insulators used in building construction are those made of fine particles or fibers with void spaces between them filled with a gas such as air.

### Convection

As a fire begins to grow, the air around it is heated by conduction. The hot air and products of combustion rise. If you hold your hand over a flame, you are able to feel the heat even though your hand is not in direct contact with the flame. The heat is being transferred to your hand by convection. *Convection* is the transfer of heat energy by the movement of heated liquids or gases. When heat is transferred by convection, there is movement or circulation of a fluid (any substance — liquid or gas — that will flow) from one place to another. As with all heat transfer, the flow of heat is from the warmer area to the cooler area (Figure 2.4).

### Radiation

If you hold your hand a few inches (millimeters) to the side of the small fire used as an example in the preceding section, you would also be able to feel heat. This heat reaches your hand by radiation. *Radiation* is the transmission of energy as an electromagnetic wave (such as light waves, radio waves, or X rays) without an intervening medium. Because it is an electromagnetic wave, the

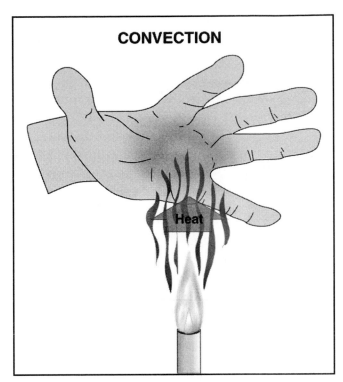

**Figure 2.4** Convection is the transfer of heat energy by the movement of heated liquids or gases.

energy travels in a straight line at the speed of light. All warm objects radiate heat. The best example of heat transfer by radiation is the sun's heat. The energy travels at the speed of light from the sun through space (a vacuum) and warms the earth's surface. Radiation is the cause of many exposure fires (fires ignited in fuel packages or buildings that are remote from the fuel package or building of origin). As a fire grows, more and more energy radiates from it in the form of heat. In large fires, it is possible for the radiated heat to cause ignition of buildings or other fuel packages some distance away (Figure 2.5). Heat energy being transmitted by radiation travels through vacuums and substantial air spaces that would normally disrupt conduction and convection. Materials that reflect radiated energy disrupt the transmission of heat.

## Matter

As you look at the world around you, the physical materials you see are called matter. It is said that matter is the "stuff" that makes up our universe. *Matter* is anything that occupies space and has mass. Matter can be described by its physical appearance or more technically by its physical properties such as mass, size, or volume.

**Figure 2.5** Radiated heat is one of the major sources of fire spread to exposures.

In addition to those properties that can be measured, matter also possesses properties that can be observed such as its physical state (solid, liquid, or gas), color, or smell. One of the best and most common examples of the physical states of matter is water. At normal atmospheric pressure (the pressure exerted by our atmosphere on all objects) and temperatures above 32°F (0°C), water is found as a liquid. At sea level, *atmospheric pressure* is defined as 760 mm of mercury measured on a barometer. When the temperature of water falls below 32°F (0°C) and the pressure remains the same, water changes state and becomes a solid called ice. At temperatures above its boiling point, water changes state to a gas called steam.

Temperature, however, is not the only factor that determines when a change of state occurs. The other factor is pressure. As previously stated, atmospheric pressure at sea level is 760 mm of mercury. Temperatures given for the freezing and boiling points of substances are normally based on this pressure. As the pressure on the surface of a substance decreases, so does the temperature at which it boils. The opposite is also true. If the pressure on the surface increases, so will the boiling point. This is the principle used in pressure cookers. The boiling point of the liquid increases as the pressure inside the vessel increases. Thus, foods cook faster in the device because the temperature of the boiling water is greater than 212°F (100°C).

Matter can also be described using terms derived from its physical properties of mass and volume. For solids, *density* is a measure of how tightly the molecules of a substance are packed together. Density is determined by dividing the mass of a substance by its volume. It is expressed as $kg/m^3$ in SI and $lb/ft^3$ in the Customary System (Figure 2.6).

For liquids, the common description used is specific gravity. *Specific gravity* is the ratio of the mass of a given volume of a liquid compared with the

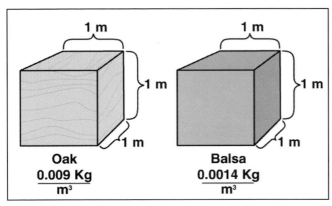

**Figure 2.6** The differences in density are reflected in the differences in weight per unit.

mass of an equal volume of water. Thus, water has a specific gravity of 1. Liquids with a specific gravity less than 1 are lighter than water, while those with a specific gravity greater than 1 are heavier than water.

For gases, the description used is vapor density. *Vapor density* is defined as the density of gas or vapor in relation to air. Because air is used for the comparison, it has a vapor density of 1 (as with specific gravity and liquids). Gases with a vapor density of less than 1 will rise, and those with vapor densities greater than 1 will fall to the surface.

## Conservation of Mass and Energy

As fire consumes a fuel, the mass of the fuel is reduced. What happens to this material? Where does it go? The answer to these questions is one of the basic concepts of modern physical science: the *Law of Conservation of Mass-Energy* (commonly shortened to the *Law of Conservation of Mass*). The law states: *Mass and energy may be converted from one to another, but there is never any net loss of total mass-energy.* In other words, mass and energy are neither created nor destroyed. The law is fundamental to the study of the science of fire. The reduction in the mass of a fuel results in the release of energy in the form of light and heat. Researchers use this principle to determine the heat release rate of materials with instruments that determine mass loss and temperature gain when a fuel is burned.

The fire investigator applies this concept as he views a fire scene and determines the fuel packages that were consumed by the fire. As the volume of fuel available to burn increases, so does the damage that results from the fire.

## Chemical Reactions

Before we begin the discussion of combustion and fire growth, it is important to understand the concept of chemical reactions. Whenever matter is transformed from one state to another or a new substance is produced, chemists describe the transformation as a *chemical reaction*. The simplest of these reactions occurs when matter changes state, which is called a *physical change*. In a physical change, the chemical makeup of the substance is not altered. The change of state that occurs when water freezes is a physical change.

A more complex reaction occurs when substances are transformed into new substances with different physical and chemical properties. These changes are defined as *chemical changes*. The change that occurs when hydrogen and oxygen are combined to form water is a chemical change. In this case, the chemical and physical properties of the materials being combined are altered. Two materials that are normally gases at room temperature are converted into a substance that is a clear liquid at room temperature.

Chemical and physical changes almost always involve an exchange of energy. Reactions that give off energy as they occur are called *exothermic*. Reactions that absorb energy as they occur are called *endothermic*. When fuels are burned in air, the fuel vapors chemically react with the oxygen in the air, and both heat and light energies are released in an exothermic reaction. Water changing state from liquid to gas (steam) requires the input of energy, thus the conversion is endothermic.

One of the more common chemical reactions on earth is oxidation. *Oxidation* is the formation of a chemical bond between oxygen and another element. Oxygen is one of the more common elements on earth (our atmosphere is composed of approximately 21 percent oxygen) and reacts with almost every other element found on the planet. The oxidation reaction releases energy or is exothermic. The most familiar example of an oxidation reaction is rusting of iron. The combination of oxygen and iron produces a flaky red compound called iron oxide or more commonly, rust. Because this is an exothermic process, it always produces heat. Normally the process is very slow, and the heat dissipates before it is noticed.

However, if the material that is rusting is in a confined space and the heat is not allowed to dissipate, the oxidation process will cause the temperature in the space to increase. One of the most common examples of heat production in confined spaces is in cargo ships loaded with iron filings. Oxidization of the filings confined within the hold of the ship generates heat that cannot be dissipated because of its location This heat is conducted to the hull and subsequently to the water outside the ship. When the vessel is in motion, the heat is transferred to the water the ship is moving through and goes unnoticed. When the ship is stationary, however, the fact that heat is being conducted to the surrounding water becomes apparent when the water near the ship begins to boil. While the temperature does not usually increase to the point that flaming ignition (fire) occurs, the condition can be quite dramatic.

 ## Combustion

Fire and combustion are terms that are often used interchangeably. Technically, however, fire is a form of combustion. The *Principles of Fire Protection Chemistry* (written by Richard L. Tuve) defines *combustion* as "a self-sustaining chemical reaction yielding energy or products that cause further reactions of the same kind." Combustion, using the term discussed earlier, is an exothermic reaction. Also, the *Principles of Fire Protection Chemistry* defines *fire* as "a rapid, self-sustaining oxidization process accompanied by the evolution of heat and light of varying intensities." The time it takes a reaction to occur is the determining factor in the type of reaction that is observed. At the very slow end of the time spectrum is rusting, where the reaction is too slow to be observed. At the upper end of the spectrum are explosions that result from the very rapid reaction of a fuel and an oxidizer. In these cases, a large amount of energy is released over a very short period of time (Figure 2.7).

### Fire Tetrahedron

For many years, the fire triangle (oxygen, fuel, and heat) was used to teach the components of fire. While this simple example is useful, it is not technically correct. Four components are necessary for combustion to occur:

- Oxygen (oxidizing agent)
- Fuel
- Heat
- Self-sustained chemical reaction

These components can be graphically described as the *fire tetrahedron* (Figure 2.8). Each compo-

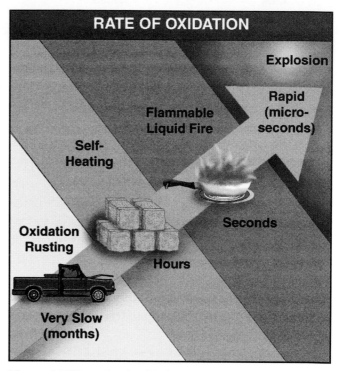

**Figure 2.7** The rate of oxidation varies from very slow to rapid.

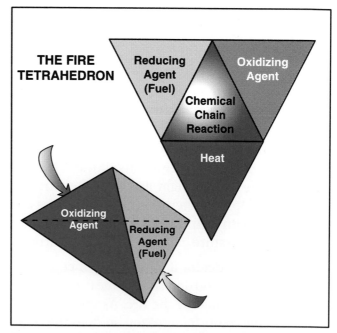

**Figure 2.8** The components of the fire tetrahedron.

nent of the tetrahedron must be in place for combustion to occur. This concept is extremely important to students of fire suppression, prevention, and investigation. Remove any one of the four components, and combustion will not occur. If ignition has already occurred, the fire is extinguished when one of the components is removed from the reaction.

To better understand fire and its behavior, each of the components of the tetrahedron is discussed in the following sections.

### Oxidizing Agent

*Oxidizing agents* are those materials that yield oxygen or other oxidizing gases during the course of a chemical reaction. Oxidizers are not themselves combustible, but they support combustion when combined with a fuel. While oxygen is the most common oxidizer, there are other substances that fall into the category. Table 2.3 lists other common oxidizers.

For the purposes of this discussion, the oxygen in the air around us is considered the primary oxidizing agent. Normally, air consists of about 21 percent oxygen. At room temperature (70°F or 21°C), combustion is supported at oxygen concentrations as low as 14 percent. Research shows, however, that as temperatures in a compartment fire increase, lower concentrations of oxygen are needed to support flaming combustion. In studies of compartment fires, flaming combustion has been observed at post-flashover temperature conditions (known as the fully developed and decay stages) when oxygen concentrations have been very low (see the Fire Development section). Some research indicates the concentration can be less than 2 percent. The depletion of oxygen in a fire also results in a significant life-safety hazard. Table 2.4 lists the physiological effects of reduced oxygen on humans.

When oxygen concentrations exceed 21 percent, the atmosphere is said to be oxygen enriched. Under these conditions, materials exhibit very different burning characteristics. Materials that burn at normal oxygen levels burn more rapidly and may ignite much easier than normal. Some petroleum-based materials autoignite in oxygen-enriched atmospheres. Many materials that do not burn at normal oxygen levels burn readily in oxygen-enriched atmospheres. One such material is Nomex· fire-resistant material, which is used to construct much of the protective clothing worn by firefighters. At normal oxygen levels, Nomex· does not burn. When placed in an oxygen-enriched atmosphere of

| Table 2.3 Common Oxidizers | |
|---|---|
| Bromates | Bromine |
| Chlorates | Chlorine |
| Fluorine | Iodine |
| Nitrates | Nitric acid |
| Nitrites | Perchlorates |
| Permanganates | Peroxides |

| Table 2.4 Physiological Effects of Reduced Oxygen (Hypoxia) | |
|---|---|
| Oxygen in Air (Percent) | Symptoms |
| 21 | None — normal conditions |
| 17 | Some impairment of muscular coordination; increase in respiratory rate to compensate for lower oxygen content |
| 12 | Dizziness, headache, rapid fatigue |
| 9 | Unconsciousness |
| 6 | Death within a few minutes from respiratory failure and concurrent heart failure |

**NOTE:** These data cannot be considered absolute because they do not account for difference in breathing rate or length of time exposed.

These symptoms occur only from reduced oxygen. If the atmosphere is contaminated with toxic gases, other symptoms may develop.

approximately 31 percent oxygen, however, Nomex ignites and burns vigorously. These conditions can be found in health care facilities, industrial occupancies, and even private homes where occupants use liquid oxygen breathing equipment. Investigators must keep these situations in mind as they analyze fire scenes that may involve oxygen-enriched atmospheres because the fire damage in the areas will be significantly greater than in those with a normal oxygen concentration.

## Fuel

*Fuel* is the material or substance being oxidized or burned in the combustion process. In scientific terms, the fuel in a combustion reaction is known as the *reducing agent.* Most common fuels contain carbon along with combinations of hydrogen and oxygen. These fuels can be further broken down into hydrocarbon-based fuels (such as gasoline, fuel oil, and plastics) and cellulose-based materials (such as wood and paper). There are also other fuels that are less complex in their chemical makeup, including hydrogen gas and combustible metals such as magnesium and sodium. In the combustion process, there are two key fuel-related factors: the physical state of the fuel and its distribution. These factors are discussed in the following paragraphs.

From the earlier discussion on matter it should be understood that a fuel may be found in any of three states of matter: solid, liquid, or gas (Figure 2.9). To burn, however, fuels must normally be in the gaseous state. For solids and liquids, energy must be expended to cause these state changes.

Fuel gases are evolved from solid fuels by pyrolysis. *Pyrolysis* is the chemical decomposition of a substance through the action of heat. Simply stated, as solid fuels are heated, combustible materials are driven from the substance. If there is sufficient fuel and heat, the process of pyrolysis generates sufficient quantities of burnable gases to ignite if the other elements of the fire tetrahedron are present.

Because of their nature, solid fuels have a definite shape and size. This property significantly affects the ease of ignition of a fuel. Of primary consideration is the surface-to-mass ratio of the fuel. The *surface-to-mass ratio* is the surface area of the fuel in relation to the mass. One of the best

**Figure 2.9** Fuels exist in one of three states of matter.

examples of the surface-to-mass ratio is wood. To produce usable materials, a tree must be cut into a log. The mass of this log is very high, but the surface area is relatively low, thus the surface-to-mass ratio is low. The log is then milled into boards. The result of this process is to reduce the mass of the individual boards as compared to the log, but the resulting surface area is increased thus increasing the surface-to-mass ratio. The sawdust that is produced as the lumber is milled has an even higher surface-to-mass ratio. If the boards are sanded, the resulting dust has the highest surface-to-mass ratio of any of the examples. As this ratio increases, the fuel particles become smaller, more finely divided, (for example, sawdust as opposed to logs), and the ignitability increases tremendously (Figure 2.10). As the surface area increases, more of the material is exposed to the heat and thus generates more burnable gases due to pyrolysis.

A solid fuel's actual position also affects the way it burns. If the solid fuel is in a vertical position, fire spread will be more rapid than if it is in a horizontal position. For example, if you were to ignite a sheet of 1/8-inch plywood paneling that was lying horizontally on two sawhorses, the fire would consume the fuel at a relatively slow rate. The same type of paneling in the vertical position would burn much more rapidly. The rapidity of fire spread is due to increased heat transfer through convection as well as conduction and radiation.

For liquids, fuel gases are generated by a process called vaporization. In scientific terms, *vaporiza-* *tion* is the transformation of a liquid to its vapor or gaseous state. The transformation from liquid to vapor or gas occurs as molecules of the substance escape from the liquid's surface into the surrounding atmosphere. In order for the molecules to break free of the liquid surface, there must be some energy input. In most cases, this energy is provided in the form of heat. For example, water left in a pan eventually evaporates. The energy required for this process comes from the sun or surrounding environment. Water in the same pan placed on a stove and heated to boiling vaporizes much more rapidly because there is more energy being applied. The rate of vaporization is determined by the substance and the amount of heat energy applied to it.

Vaporization of liquid fuels generally requires less energy input than does pyrolysis for solid fuels. This is primarily caused by the different densities of substances in solid and liquid states and by the fact that molecules of a substance in the liquid state have more energy than when they are in the solid state. Solids also absorb more of the energy because of their mass. The volatility or ease with which a liquid gives off vapor influences its ignitability. All liquids give off vapors to a greater or lesser degree in the form of simple evaporation. Liquids that easily give off quantities of flammable or combustible vapors can be dangerous.

Like the surface-to-mass ratio for solid fuels, the surface-to-volume ratio of liquids is an important factor in their ignitability. A liquid assumes the shape of its container. Thus, when a spill or release occurs, the liquid assumes the shape of the ground (flat), flows, and accumulates in low areas. When contained, the specific volume of a liquid has a relatively low surface-to-volume ratio. When it is released, this ratio increases significantly as does the amount of fuel vaporized from the surface.

Gaseous fuels can be the most dangerous of all fuel types because they are already in the natural state required for ignition. No pyrolysis or vaporization is needed to ready the fuel, and less energy is required for ignition.

For combustion to occur after a fuel has been converted into a gaseous state, the fuel must be mixed with air (oxidizer) in the proper ratio. The

**Figure 2.10** Materials with a high surface-to-mass ratio require less energy to ignite.

range of concentrations of the fuel vapor and air (oxidizer) is called the *flammable (explosive) range*. The flammable range of a fuel is reported using the percent by volume of gas or vapor in air for the *lower flammable limit* (LFL) and for the *upper flammable limit* (UFL). The lower flammable limit is the minimum concentration of fuel vapor and air that supports combustion. Concentrations that are below the LFL are said to be *too lean* to burn. The *upper flammable limit* is the concentration above which combustion cannot take place. Concentrations that are above the UFL are said to be *too rich* to burn.

Table 2.5 presents the flammable ranges for some common materials. The flammable limits for combustible gases are presented in chemical handbooks and documents such as National Fire Protection Association (NFPA) Standard 49, *Hazardous Chemicals Data*, and NFPA 325, *Fire Hazard Properties of Flammable Liquids, Gases, and Volatile Solids*. The limits are normally reported at ambient temperatures and atmospheric pressures. Variations in temperature and pressure can cause the flammable range to vary considerably. Generally, increases in temperature or pressure broaden the range and decreases narrow it.

### Table 2.5
### Flammable Ranges for Selected Materials

| Material | Lower Flammable Limit (LFL) | Upper Flammable Limit (UFL) |
|---|---|---|
| Acetylene | 2.5 | 100.0 |
| Carbon Monoxide | 12.5 | 74.0 |
| Ethyl Alcohol | 3.3 | 19.0 |
| Fuel Oil No. 1 | 0.7 | 5.0 |
| Gasoline | 1.4 | 7.6 |
| Hydrogen | 4.0 | 75.0 |
| Methane | 5.0 | 15.0 |
| Propane | 2.1 | 9.5 |

*Source: NFPA 325, Fire Hazard Properties of Flammable Liquids, Gases, and Volatile Solids, 1994 edition.*

**Fire Loads**

It is often helpful to the investigator to identify groups of fuels or fuel packages that were present in an area or building compartment involved in a fire. In outside areas, fuel packages could be clusters of underbrush or trees growing close together. In an outside storage area, a fuel package might be a tank of fuel oil or flammable liquid or a pile of lumber or building materials (Figure 2.11). In buildings, combustible components (both structural and interior finish) and the contents can be considered fuel packages (Figure 2.12). A foam-padded upholstered chair or couch in a living room, a mattress and box spring unit in a bedroom, or a computer and office furniture in a business office would be considered fuel packages.

The total amount (mass) of fuel in a compartment or specific location multiplied by the heat of combustion of the materials is called the *fuel load* or *fire load*. The term is commonly used to describe the maximum heat that would be released if all the materials in an area or compartment burned. The concept of fire load was the basis for many of the fire-resistance requirements found in today's building codes. In one study, researchers found the fire loads of typical basement recreation rooms to average about 5.8 pounds per square foot (psf) or 28.3 kg/m$^2$. Fire loading is normally expressed in terms of the heat of combustion of wood. Materials with different heats of combustion are converted to be equivalent to wood. The available fuel in a space and the proximity of fuel packages to each other have a significant impact on the growth and development of fires. As the amount of available fuel increases, the potential heat release rate of a fire in the compartment increases. If fuel packages are very close together, the amount of energy required for a fire in one package to generate enough heat to ignite the nearby (target) fuel package is less than that for target packages at greater distances.

### Heat

*Heat* is the energy component of the fire tetrahedron. When heat comes into contact with a fuel, the energy supports the combustion reaction in the following ways:

- By causing the pyrolysis of solid fuels (Figure 2.13)

- By causing vaporization of liquid fuels (Figure 2.14)

- By providing the energy necessary for ignition to occur

**Figure 2.11** Flammable liquid storage tanks are a form of fuel package.

**Figure 2.12** The lumber stacked in front of this building as well as the structural components of the building are examples of fuel packages.

- By causing the continuous production and ignition of fuel vapors or gases so that the combustion reaction can continue

Most of the energy types discussed earlier in the chapter produce heat. For our discussion of fire and its behavior, however, chemical, electrical, and mechanical energy are the most common sources of heat that result in the ignition of a fuel. Each of these sources is discussed in depth in this section.

*Chemical.* Chemical heat energy is the most common source of heat in combustion reactions. When any combustible is in contact with oxygen, oxidation occurs. This process almost always results in the production of heat. The heat generated when a common match burns is an example of chemical heat energy.

*Self-heating* (also known as spontaneous heating) is a form of chemical heat energy that occurs when a material increases in temperature without the addition of external heat. Normally the heat is produced slowly by oxidation and is lost to the surroundings almost as fast as it is generated. In

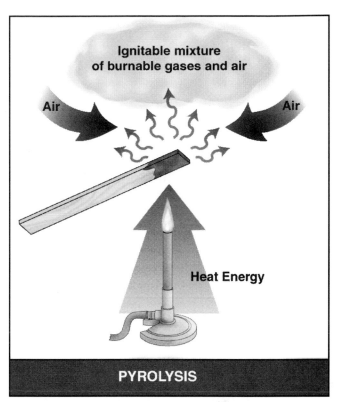

**PYROLYSIS**

**Figure 2.13** Pyrolysis takes place as the wood decomposes from the action of the heat, generating vapors.

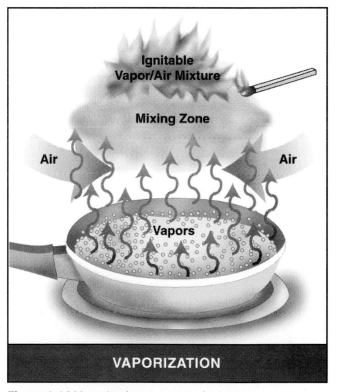

**VAPORIZATION**

**Figure 2.14** Vaporization occurs as fuel gases are generated from the action of heat.

order for self-heating to progress to spontaneous ignition, the material must be heated to its ignition temperature (minimum temperature at which self-sustained combustion occurs for a specific substance). For spontaneous ignition to occur, the following events must occur:

- The rate of heat production must be great enough to raise the temperature of the material to its ignition temperature.

- The available air supply (ventilation) in and around the material being heated must be adequate to support combustion.

- The insulation properties of the material immediately surrounding the fuel must be such that the heat being generated does not dissipate.

An example of a situation that could lead to spontaneous ignition would be a number of rags that are soaked with boiled linseed oil and that are rolled into a ball and thrown into a corner. If the heat that is generated by the natural oxidation of the oil and cloth is not allowed to dissipate, either by movement of air around the rags or some other method of heat transfer, the temperature of the cloth will eventually become sufficient to cause ignition (Figure 2.15).

The rate of the oxidation reaction, and thus the heat production, increases as more heat is generated and held by the materials insulating the fuel. In fact, the rate at which most chemical reactions occur doubles with each 18°F (10°C) increase in the temperature of the reacting materials. The more heat generated and absorbed by the fuel, the faster the reaction causing the heat generation. When the heat generated by a self-heating reaction exceeds the heat being lost, the material may reach its ignition temperature and ignite spontaneously. Table 2.6 lists some common materials that are subject to self-heating.

A common cause of self-heating in agricultural products is the result of the generation of heat by microbiological activity. This activity is the decomposition of the material by bacteria. In these materials, the decomposition action is increased when they are wet. The bacterial action causes the temperatures in the material to increase. This temperature increase actually causes the death of the organisms when the temperatures exceed 160° to 175°F ( 70° C to 80°C). At this point, any continued heating of the material is the result of the oxidation process that the microbiological activity originated. As previously discussed, the rate of the reaction doubles with each 18°F (10°C) increase in the temperature.

*Electrical.* Electrical heat energy has the ability to generate high temperatures that are capable of igniting any combustible materials near the heated

### Table 2.6
### Materials Subject to Spontaneous Heating

| Material | Tendency |
|---|---|
| Charcoal | High |
| Fish meal/fish oil | High |
| Linseed oil rags | High |
| Brewers grains/feed | Moderate |
| Fertilizers | Moderate |
| Foam rubber | Moderate |
| Hay | Moderate |
| Manure | Moderate |
| Iron metal powder | Moderate |
| Waste paper | Moderate |
| Rags (bales) | Low to moderate |

*Source: Fire Protection Handbook,* NFPA 18th edition, Table A-10, 1997.

PRE-FLASHOVER CONDITION

**Figure 2.15** Heat impacting the surface of the fuel drives off additional vapors/gases.

area. Electrical heating can occur in several ways, including the following (Figures 2.16 a and b):

- Current flow through a resistance
- Overcurrent or overload
- Arcing
- Sparking
- Static
- Lightning

***Mechanical.*** Mechanical heat energy is generated by friction and compression. *Heat of friction* is created by the movement of two surfaces against each other. This movement results in heat and/or sparks being generated. *Heat of compression* is generated when a gas is compressed. Diesel engines ignite fuel vapor without a spark plug by the use of this principle. It is also the reason that self-contained breathing apparatus (SCBA) bottles feel warm to the touch after they have been filled (Figure 2.17).

***Nuclear.*** Nuclear heat energy is generated when atoms are either split apart (fission) or combined (fusion). In a controlled setting, fission is used to heat water to drive steam turbines and produce

**Figure 2.16a** An example of a common electrical ignition source. *Courtesy of R.P. Rohten.*

**Figure 2.16b** Illegal wiring is a common electrical ignition source. *Courtesy of R.P. Rhoten.*

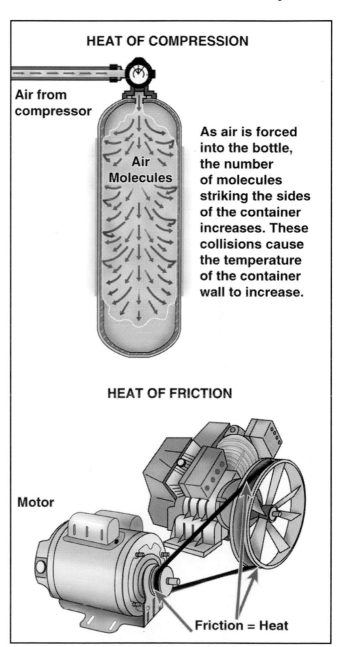

**HEAT OF COMPRESSION**

Air from compressor

Air Molecules

As air is forced into the bottle, the number of molecules striking the sides of the container increases. These collisions cause the temperature of the container wall to increase.

**HEAT OF FRICTION**

Motor

Friction = Heat

**Figure 2.17** Examples of mechanical heat energy.

electricity. Fusion reactions cannot be contained at this time and have no commercial use. The sun's heat (solar energy) is a product of a fusion reaction and thus is a form of nuclear energy.

## Self-Sustained Chemical Reaction

Combustion is a complex reaction that requires a fuel (in the gaseous or vapor state), an oxidizer, and heat energy to come together in a very specific way. Once flaming combustion or fire occurs, it can only continue when enough heat energy is produced to cause the continued development of fuel vapors or gases. Scientists call this type of reaction a chain reaction. A *chain reaction* is a series of reactions that occur in sequence with the results of each individual reaction being added to the rest. An excellent illustration of a chain reaction is given by Faughn, Chang, and Turk in their textbook *Physical Science*:

*An example of a chemical chain reaction is a forest fire. The heat from one tree may initiate the reaction (burning) of a second tree, which, in turn ignites a third, and so on. The fire will then go on at a steady rate. But if one burning tree ignites, say, two others, and each of these two ignite two more, for a total of four, and so on, the rate of burning speeds rapidly. Such uncontrolled, runaway chain reactions are at the heart of nuclear bombs.*

*The self-sustained chemical reaction and the related rapid growth are the factors that separate fire from slower oxidation reactions. Slow oxidation reactions do not produce heat fast enough to reach ignition and never generate sufficient heat to become self-sustained. Examples of slow oxidation are rusting of iron (given earlier) and yellowing of paper.*

## Fire Development

When the four components of the fire tetrahedron come together, ignition occurs. For a fire to grow beyond the first material ignited, heat must be transmitted beyond the first material to additional fuel packages. In the early development of a fire, heat rises and forms a plume of hot gas. If a fire is in the open (outside or in a large building), the fire plume rises unobstructed, and air is drawn (entrained) into it as it rises. Because the air being pulled into the plume is cooler than the fire gases, this action has a cooling effect on the gases above the fire. The spread of fire in an open area is primarily due to heat energy that is transmitted from the plume to nearby fuels. Fire spread in outside fires can be increased by wind and sloping terrain that allow exposed fuels to be preheated (Figure 2.18).

The development of fires in a compartment is more complex than those in the open. For the pur-

**Figure 2.18** Several factors influence fire development in the open.

poses of this discussion, a *compartment* is an enclosed room or space within a building. The term *compartment fire* is defined as a fire that occurs within such a space. The growth and development of a compartment fire is usually controlled by the availability of fuel and oxygen. When the amount of fuel available to burn is limited, the fire is said to be *fuel controlled*. When the amount of available oxygen is limited, the condition is called *ventilation controlled*.

Recently, researchers have attempted to describe compartment fires in terms of stages or phases that occur as the fire develops. These stages are as follows:

- Ignition

- Growth

- Flashover

- Fully developed

- Decay

Figure 2.19 shows the development of a compartment fire in terms of time and temperature. It should be noted that the stages are an attempt to describe the complex reaction that is occurring as a fire develops in a space with no suppression action taken. The ignition and development of a compartment fire are very complex and influenced by many variables. As a result, all fires may not develop through each of the stages described. The information is presented to depict fire as a dynamic event that is dependent on many factors for its growth and development.

## Ignition

Ignition describes the period when the four elements of the fire tetrahedron come together and combustion begins. The physical act of ignition can be *piloted* (caused by a spark or flame) or *nonpiloted* (caused when a material reaches its ignition temperature as the result of self-heating) such as spontaneous ignition. At this point, the fire is small and generally confined to the material (fuel) first ignited. All fires — in an open area or within a compartment — occur as a result of some type of ignition.

## Growth

Shortly after ignition, a fire plume begins to form above the burning fuel. As the plume develops, it begins to draw or entrain air from the surrounding space into the column. The initial growth is similar to that of an outside unconfined fire, with the growth a function of the fuel first ignited. Unlike an unconfined fire, the plume in a compartment is rapidly affected by the ceiling and walls of the space. The first impact is the amount of air that is entrained into the plume. Because the air is cooler than the hot gases generated by the fire, the air has a cooling effect on the temperatures within the plume. The location of the fuel package in relation to the compartment walls determines the amount of air that is entrained and thus the amount of cooling that takes place. Fuel packages near walls entrain less air and thus have higher plume temperatures. Fuel packages in corners entrain even less air and have the highest plume temperatures. This factor significantly affects the temperatures in the developing

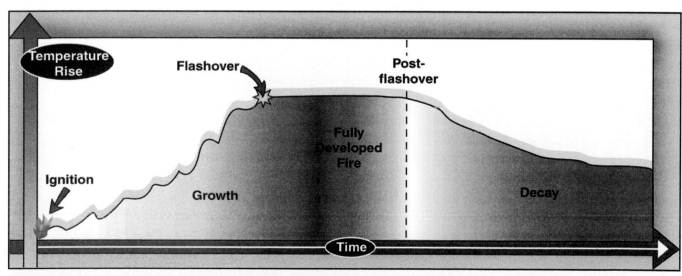

**Figure 2.19** The development of a compartment fire in terms of time and temperature.

hot-gas layer above the fire. As the hot gases rise, they begin to spread outward when they hit the ceiling. The gases continue to spread until they reach the walls of the compartment. The depth of the gas layer then begins to increase.

The temperatures in the compartment during this period depend on the amount of heat conducted into the compartment ceiling and walls as the gases flow over them and on the location of the initial fuel package and the resulting air entrainment. Research shows that the gas temperatures decrease as the distance from the centerline of the plume increases. Figure 2.20 shows the plume in a typical compartment fire and the factors that impact the temperature of the developing hot-gas layer.

The growth stage continues if enough fuel and oxygen are available. Compartment fires in the growth stage are generally fuel controlled. As the fire grows, the overall temperature in the compartment increases as does the temperature of the gas layer at the ceiling level.

### Flashover

*Flashover* is the transition between the growth and the fully developed fire stage and is not a specific event such as ignition. During flashover, conditions in the compartment change very rapidly as the fire changes from one that is dominated by the burning of the materials first ignited to one that involves all the exposed combustible surfaces within the compartment. The hot-gas layer that develops at the ceiling level during the growth stage causes radiant heating of combustible materials remote from the origin of the fire. Typically, radiant energy (heat flux) from the hot-gas layer exceeds 20 kW/m² when flashover occurs. This radiant heating causes pyrolysis in the combustible materials in the compartment. The gases generated during this time are heated to their ignition temperature by the radiant energy from the gas layer at the ceiling.

While scientists define flashover in many ways, most base their definition on the temperature in a compartment that results in the simultaneous ignition of all the combustible contents in the space. While no exact temperature is associated with this occurrence, a range from approximately 900°F to 1,200°F (500°C to 650°C) is widely used. This range correlates with the ignition temperature of carbon monoxide (CO) (1,128°F or 609°C), one of the most common gases given off from pyrolysis.

Just prior to flashover, several things are happening within the burning compartment: The temperatures are rapidly increasing, additional fuel packages are becoming involved, and the fuel packages in the compartment are giving off combustible gases as a result of pyrolysis. As flashover occurs, the combustible materials in the compartment and the gases given off from pyrolysis ignite. The result is full-room involvement. The heat release from a fully developed room at flashover can be on the order of 10,000 kW or more.

Occupants who have not escaped from a compartment before flashover occurs are not likely to survive. Firefighters who find themselves in a compartment at flashover are at extreme risk even while wearing their personal protective equipment.

### Fully Developed Fire

The fully developed fire stage occurs when all combustible materials in the compartment are involved

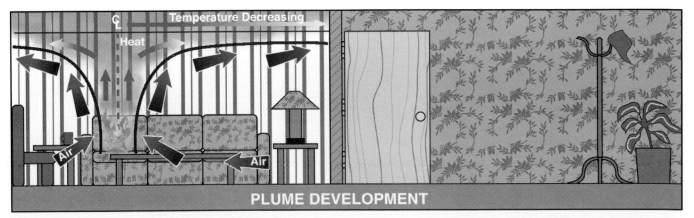

**Figure 2.20** Initially, the temperature of the fire gases decreases as they move away from the centerline of the plume.

in fire (Figure 2.21). During this period of time, the burning fuels in the compartment are releasing the maximum amount of heat possible for the available fuel packages and are producing large volumes of fire gases. The heat released and the volume of fire gases produced depend on the number and size of the ventilation openings in the compartment. The fire frequently becomes ventilation controlled, and thus large volumes of unburned gases are produced. During this stage, hot unburned fire gases are likely to begin flowing from the compartment of origin into adjacent spaces or compartments. These gases ignite as they enter a space where air is more abundant.

### Decay

As the fire consumes the available fuel in the compartment, the rate of heat release begins to decline. Once again the fire becomes fuel controlled, the amount of fire diminishes, and the temperatures within the compartment begin to decline. The remaining mass of glowing embers can, however, result in moderately high temperatures in the compartment for some time.

### Factors That Impact Fire Development

As the fire progresses from ignition to decay, several factors affect its behavior and development within the compartment:

- Size, number, and arrangement of ventilation openings
- Volume of the compartment
- Thermal properties of the compartment enclosures
- Ceiling height of the compartment

- Size, composition, and location of the fuel package that is first ignited
- Availability and locations of additional fuel packages (target fuels)

For a fire to develop, enough air to support burning beyond the ignition stage must be available. The size and number of ventilation openings in a compartment determine how the fire develops within the space. The compartment size and shape and the ceiling height determine whether a significant hot-gas layer will form. The location of the initial fuel package is also very important in the development of the hot-gas layer. The plumes of burning fuel packages in the center of a compartment entrain more air and are cooler than those against walls or in corners of the compartment.

The temperatures that develop in a burning compartment are the direct result of the energy released as the fuels burn. Because matter and energy are conserved, any loss in mass caused by the fire is converted to energy. In a fire, the resulting energy is in the form of heat and light. The amount of heat energy released over time in a fire is called the *heat release rate* (HRR). HRR is measured in Btu/s or kilowatts (kW). The heat release rate is directly related to the amount of fuel being consumed over time and the heat of combustion (the amount of heat a specific mass of the substance gives off when burned) of the fuel being burned. See Table 2.7 for maximum heat release rates for several common items. This information gives representative numbers for typical fuel items.

Fire investigators should be able to recognize the damage from fuel packages in a building or compartment and use this information as part of the

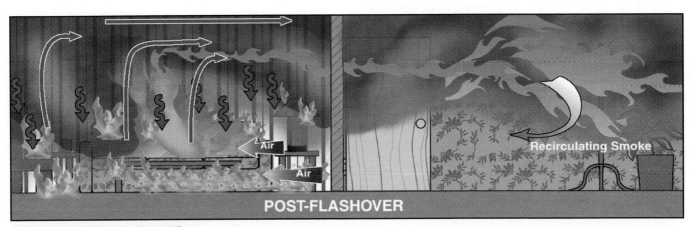

**Figure 2.21** A fully developed fire.

## Table 2.7
## Heat Release Rates for Common Materials

| Material | Maximum Heat Release Rate | |
|---|---|---|
| | kW | Btu/s |
| Wastebasket (0.53 kg) with milk cartons (0.40 kg) | 15 | 14.2 |
| Upholstered chair (cotton padded) (31.9 kg) Four stacking chairs | 370 | 350.7 |
| (metal frame, polyurethane foam padding) (7.5 kg each) | 160 | 151.7 |
| Upholstered chair (polyurethane foam) (28.3 kg) | 2,100 | 1,990.0 |
| Mattress (cotton and jute) (25 kg) | 40 | 37.9 |
| Mattress (polyurethane foam) (14 kg) | 2,630 | 2,492.9 |
| Mattress and box springs (cotton and polyurethane foam) (62.4 kg) | 660 | 626.0 |
| Upholstered sofa (polyurethane foam) (51.5 kg) | 3,200 | 3,033.0 |
| Gasoline/kerosene (2 sq ft pool) | 400 | 379.0 |
| Christmas tree (dry) (7.4 kg) | 500 | 474.0 |

*Source: NFPA 921, Guide for Fire and Explosion Investigations, Table 3-4; NBSIR 85-3223 Data Sources for Parameters Used in Predictive Modeling of Fire Growth and Smoke Spread; and NBS Monograph 173, Fire Behavior of Upholstered Furniture.*

analysis of the fire scene. Materials with high heat release rates such as polyurethane-foam-padded furniture, polyurethane-foam mattresses, or stacks of wooden pallets, for example, would be expected to burn rapidly once ignition occurs. Fires in materials with lower heat release rates would be expected to take longer to develop. In general, low-density materials (such as polyurethane foam) burn faster (have a higher HRR) than higher-density materials (cotton padding) of similar makeup.

One final relationship between the heat generated in a fire and fuel packages is the ignition of fuel packages that are remote from the first package ignited. The heat generated in a compartment fire is transmitted from the initial fuel package to other fuels in the space by all three modes of heat transfer. The heat rising in the initial fire plume is transported by convection. As the hot gases travel over surfaces of other fuels in the compartment, heat is transferred to them by conduction. Radiation plays a significant role in the transition from a growing fire to a fully developed fire in a room. As the hot-gas layer forms at the ceiling, hot particles in the smoke begin to radiate energy to the other fuel packages in the compartment. These remote fuel packages are sometimes called *target fuels*. As the radiant energy increases, the target fuels begin the process of pyrolysis and start to give off ignitable gases. When the temperature in the compartment reaches the ignition temperature of these gases, the entire room becomes involved in fire (flashover).

 ## Special Considerations

Several conditions or situations that occur during a fire's growth and development should be discussed. This section provides an overview of these conditions and the potential damage that may result during the fire.

### Flameover/Rollover

The terms *flameover* and *rollover* describe a condition where flames move through or across the unburned gases during a fire's progression. Flameover is distinguished from flashover by its involvement of only the fire gases and not the surfaces of other fuel packages within a compartment. This condition may occur during the growth stage as the hot-gas layer forms at the ceiling of the compartment. Flames may be observed in the layer when the combustible gases reach their ignition temperature. While the flames add to the total heat generated in the compartment, this condition is not flashover. Flameover may also be observed when unburned fire gases vent from a compartment during the growth and fully developed stages of a fire's development. As these hot gases vent from the burning compartment into the adjacent space, they mix with oxygen; if they are at their ignition temperature, flames often become visible in the layer. A flameover may not result in the ignition of target fuels within a compartment but may result in burn patterns on walls and other combustible building elements with which the gases are in direct contact.

### Thermal Layering of Gases

The *thermal layering* of gases is the tendency of gases to form into layers according to temperature.

Other terms sometimes used to describe this layering of gases by heat are *heat stratification* and *thermal balance*. The hottest gases tend to be in the top layer, while the cooler gases form the lower layers. Smoke — a heated mixture of air, gases, and particles — rises. If a hole is made in a roof, smoke will rise from the building or room to the outside. *Thermal layering* is critical to fire fighting activities. As long as the hottest air and gases are allowed to rise, the lower levels will be safer for firefighters (Figure 2.22).

This normal layering of the hottest gases to the top and out the ventilation opening can be disrupted if water is applied directly into the layer (Figure 2.23). When water is applied to the upper level of the layer, where the temperatures are highest, the rapid conversion to steam can cause

the gases to mix rapidly. This swirling mixture of smoke and steam disrupts normal thermal layering, and hot gases mix throughout the compartment. This process is sometimes referred to as *disrupting the thermal balance or creating a thermal imbalance.* Many firefighters have been burned when thermal layering was disrupted. Should this condition occur during fire fighting operations, the investigator could observe different patterns in the compartment.

## Backdraft

Firefighters operating at fires in buildings must use care when opening a building to gain entry or to provide horizontal ventilation (opening doors or windows). As the fire grows in a compartment, large volumes of hot, unburned fire gases can

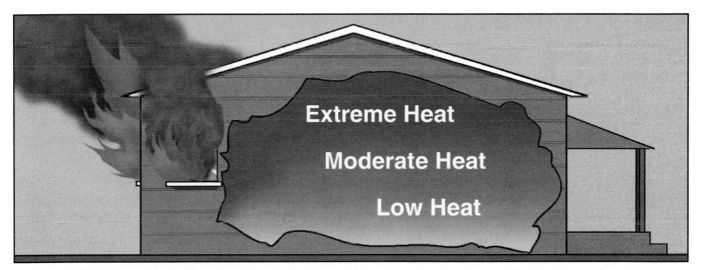

**Figure 2.22** Under normal conditions in a closed compartment, the highest levels of heat will be found near the ceiling and the lowest near the floor.

**Figure 2.23** Applying water to the upper level of the thermal layer creates a thermal imbalance.

collect in unventilated spaces. These gases may be at or above their ignition temperature but have insufficient oxygen available to actually ignite. Any action during fire fighting operations that allows air to mix with these hot gases can result in an explosive ignition called *backdraft*. Many firefighters have been killed or injured as a result of backdrafts. The potential for backdraft can be reduced with proper vertical ventilation (opening at highest point) because the unburned gases rise. Opening the building or space at the highest possible point allows them to escape before entry is made.

The following conditions may indicate the potential for a backdraft:

- Pressurized smoke exiting small openings
- Black smoke becoming dense gray-yellow
- Confinement and excessive heat
- Little or no visible flame
- Smoke leaving the building in puffs or at intervals (appearance of breathing)
- Smoke-stained windows

Should a backdraft occur, the fire investigator may observe blastlike damage in the building. Homes have been known to be moved from their foundations, and walls of multiple-story buildings have been blown out as a result of the backdraft.

## Products of Combustion

As a fuel burns, the chemical composition of the material changes (Figure 2.24). This change results in the production of new substances and the generation of energy. As a fuel is burned, some of it is actually consumed. The Law of Conservation of Mass tells us that any mass lost converts to energy. In the case of fire, this energy is in the form of light and heat. Burning also results in the generation of airborne fire gases, particles, and liquids. These materials have been referred to throughout this chapter as products of combustion or smoke. The heat generated during a fire is one of the products of combustion. In addition to being responsible for the spread of a fire, heat may also cause burns, dehydration, heat exhaustion, and injury to a person's respiratory tract.

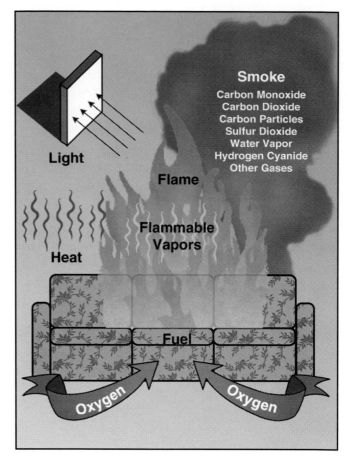

**Figure 2.24** As a fuel burns, its chemical composition changes.

While the heat energy from a fire is a danger to anyone directly exposed to it, smoke causes most deaths in fires. The materials that make up smoke vary from fuel to fuel, but generally all smoke can be considered toxic. The smoke generated in a fire contains narcotic (asphyxiant) gases and irritants. Narcotic or asphyxiant gases are those products of combustion that cause central nervous system depression that results in reduced awareness and intoxication and that can lead to loss of consciousness and death. The most common narcotic gases found in smoke are carbon monoxide (CO), hydrogen cyanide (HCN), and carbon dioxide ($CO_2$). The reduction in oxygen levels as a result of a fire in a compartment will also cause a narcotic effect in humans. Irritants found in the smoke are those substances that cause breathing discomfort (pulmonary irritants) and inflammation of the eyes, respiratory tract, and skin (sensory irritants). Depending on the fuels involved, smoke will contain numerous substances that can be considered irritants.

The most common of the hazardous substances contained in smoke is carbon monoxide. While CO is not the most dangerous of the materials found in smoke, it is almost always present when combustion occurs. While someone may be killed or injured by breathing a variety of toxic substances in smoke, carbon monoxide is the one that is most easily detected in the blood of fire victims and thus most often reported. Because the substances in smoke from compartment fires (either alone or in combination) are deadly, firefighters and investigators must use SCBA for protection when operating in smoke.

Flame is the visible, luminous body of a burning gas. When a burning gas is mixed with the proper amounts of oxygen, the flame becomes hotter and less luminous. The loss of luminosity is caused by a more complete combustion of the carbon. For these reasons, flame is considered to be a product of combustion. Of course, it is not present in those types of combustion that do not produce a flame such as smoldering fires.

 ## Conclusion

As you progress through this manual, you will find frequent references to fire behavior as part of the collection and analysis of information from the fire scene. Your knowledge of fire behavior will be used to read the patterns that remain after the fire has been extinguished. You will also use this knowledge in the development of your conclusion as to how the fire began and grew after ignition. You will use your understanding of heat release rates, as they apply to various fuels, to help you read the various patterns that are encountered in compartment fires. An understanding of fire and its behavior is fundamental to performing an investigation.

# The Basics of Building Construction as it Relates to the Investigator

**Performance Objectives**

This chapter provides information that addresses performance objectives described in NFPA 1033, *Standard for Professional Qualifications for Fire Investigator*, particularly those referenced in the following sections:

**Chapter 3  Fire Investigator**
3-2.3 a
3-2.4 b

For the fire investigator to be effective, a working knowledge of building construction techniques and materials is imperative. Investigators will use this knowledge in many ways during an investigation.

This chapter focuses on providing the fire investigator an overview of the different types of construction, construction and finish materials, and the systems that come together to make up a building. This is done to create a foundation that the fire investigator will take into the field. This foundation will become the basis for many of the judgments and decisions made during the investigation.

While the knowledge of building construction assists the fire investigator in the actual determination of the origin, cause, and spread of a fire, safety is another primary reason for this discussion. Fire investigators must be able to recognize unsafe building conditions and conduct their examinations of the building accordingly. Just as it has become recognized that there are times when it is not safe to execute an interior attack during fire suppression operations, there are times when it is not safe to allow fire investigation personnel into a severely damaged structure to examine the scene.

 ## Basic Building Classifications

A look at the different types and classifications of building construction provides the fire investigator with some insight into how buildings are constructed and how they react when they are exposed to fire. In general, construction classifications are based upon materials used in construction and the hourly fire-resistance ratings of structural components. Different building codes may use some variations in terminology. However, in the field of fire protection, buildings are grouped into five major classifications as defined in NFPA 220, *Standard on Types of Building Construction*. These classifications follow:

**NFPA 220 Classification**
**Construction Types**

| | |
|---|---|
| Type I | Fire resistive |
| Type II | Noncombustible |
| Type III | Ordinary |
| Type IV | Heavy timber |
| Type V | Wood frame |

Table 3.1 gives an overview of the fire-resistance ratings required for the individual structural components for each construction type. Each classification is designated by a three-digit number code (for example: Type I 443). The codes define the fire-resistance rating (in hours) for specific structural components, as follows:

- The first digit (4 in the example) refers to the fire-resistance rating of the exterior bearing walls.

- The second digit (4 in the example) refers to the fire-resistance rating of the structural frame including columns, beams, trusses, arches, girders, and bearing walls that support loads of more than one floor (for Type IV construction the rating is replaced with H for heavy timber).

- The third digit (3 in the example) indicates the fire-resistance rating of the floor construction (for Type IV construction the rating is replaced with H for heavy timber).

The fire-resistance ratings specified in NFPA 220 and in the model building codes are based on the ASTM E119, *Standard Test Method of Fire Tests of Building Construction.* This test is also defined by Underwriters Laboratories UL 263 and by NFPA in NFPA 251. The E119 test is the basis for most fire-resistance ratings found in the codes. This test pro-

## Table 3.1
## Fire Resistance Ratings (In Hours) For Type I Through Type V Construction

| | Type I | | Type II | | | Type III | | Type IV | Type V | |
|---|---|---|---|---|---|---|---|---|---|---|
| | 443 | 332 | 222 | 111 | 000 | 211 | 200 | 2HH | 111 | 000 |
| **Exterior Bearing Walls —** | | | | | | | | | | |
| Supporting more than one floor, columns, or other bearing walls | 4 | 3 | 2 | 1 | 0[1] | 2 | 2 | 2 | 1 | 0[1] |
| Supporting one floor only | 4 | 3 | 2 | 1 | 0[1] | 2 | 2 | 2 | 1 | 0[1] |
| Supporting a roof only | 4 | 3 | 1 | 1 | 0[1] | 2 | 2 | 2 | 1 | 0[1] |
| **Interior Bearing Walls —** | | | | | | | | | | |
| Supporting more than one floor, columns, or other bearing walls | 4 | 3 | 2 | 1 | 0 | 1 | 0 | 2 | 1 | 0 |
| Supporting one floor only | 3 | 2 | 2 | 1 | 0 | 1 | 0 | 1 | 1 | 0 |
| Supporting roofs only | 3 | 2 | 1 | 1 | 0 | 1 | 0 | 1 | 1 | 0 |
| **Columns —** | | | | | | | | | | |
| Supporting more than one floor, columns, or other bearing walls | 4 | 3 | 2 | 1 | 0 | 1 | 0 | H[2] | 1 | 0 |
| Supporting one floor only | 3 | 2 | 2 | 1 | 0 | 1 | 0 | H[2] | 1 | 0 |
| Supporting roofs only | 3 | 2 | 1 | 1 | 0 | 1 | 0 | H[2] | 1 | 0 |
| **Beams, Girders, Trusses & Arches —** | | | | | | | | | | |
| Supporting more than one floor, columns, or other bearing walls | 4 | 3 | 2 | 1 | 0 | 1 | 0 | H[2] | 1 | 0 |
| Supporting one floor only | 3 | 2 | 2 | 1 | 0 | 1 | 0 | H[2] | 1 | 0 |
| Supporting roofs only | 3 | 2 | 1 | 1 | 0 | 1 | 0 | H[2] | 1 | 0 |
| **Floor Construction** | 3 | 2 | 2 | 1 | 0 | 1 | 0 | H[2] | 1 | 0 |
| **Roof Construction** | 2 | 1½ | 1 | 1 | 0 | 1 | 0 | H[2] | 1 | 0 |
| **Exterior Nonbearing Walls**[3] | 0[1] | 0[1] | 0[1] | 0[1] | 0[1] | 0[1] | 0[1] | 0[1] | 0[1] | 0[1] |

☐ Those members that shall be permitted to be of approved combustible material.

[1] See A-3-1
[2] "H" indicates heavy timber members; see text for requirements.
[3] Exterior nonbearing walls meeting the conditions of acceptance of NFPA 285, *Standard Method of Test for the Evaluation of Flammability Characteristics of Exterior Non-Load-Bearing Wall Assemblies Containing Combustible Components Using the Intermediate-Scale, Multistory Test Apparatus,* shall be permitted to be used.

Reprinted with permission from NFPA 220, *Types of Building Construction,* Copyright © 1999, National Fire Protection Association, Quincy, MA 02269. This reprinted material is not the complete and official position of the National Fire Protection Association, on the referenced subject which is represented only by the standard in its entirety.

vides a relative measure of the performance of an assembly or product as compared to similar items. The tests are useful in that they are reproducible in the laboratory — an assembly that obtains a 1-hour rating with the E119 test from one testing laboratory should get a similar rating from any other lab doing the same test. The tests do not, however, provide specific information on how the item will perform during actual fire conditions because factors such as heat release rate of fuels, rate of temperature rise, and exposure time may differ widely from the controlled conditions. The table shows the differences between the subclassifications and where combustible materials can be used. The highest requirements for fire resistance are for Type I construction with lesser requirements for other types of construction.

Table 3.2 provides a comparison between the NFPA 220 classifications and those used by the three model building codes used in the United States.

Table 3.3 provides an overview of the construction classifications used in Canada.

### Table 3.2
### A Comparison of Construction Types (Based on the MCSC National System)

| NFPA | I (443) | I (332) | II (222) | II (111) | II (000) | III (211) | III (200) | IV (2HH) | V (111) | V (000) |
|---|---|---|---|---|---|---|---|---|---|---|
| UBC | — | I FR | II FR | II-1hr | II N | III-1hr | III N | IV HT | V 1-hr | V-N |
| BNBC | 1A | 1B | 2A | 2B | 2C | 3A | 3B | 4 | 5A | 5B |
| SBC | I | II | — | IV 1-hr | IV unp | V 1-hr | V unp | III | VI 1-hr | VI unp |

### Table 3.3
### Types of Construction and Their Firesafety Characteristics, National Building Code of Canada*

| Basic Type of Construction | Group | SubTypes | Characteristics (under fire load conditions) |
|---|---|---|---|
| Combustible Construction | I | Wood frame<br>Wood post and beam<br>Plank<br>Plastic<br>Other unprotected combustible | Fuel contributing and unstable |
| | | Heavy timber construction and other protected combustible construction | Fuel contributing but partially stable to the degree of fire resistance |
| Noncombustible construction | II | Unprotected steel construction<br>Ordinary prestressed concrete<br>Thin unprotected reinforced masonry<br>Other unprotected noncombustible construction | Nonfuel contributing but unstable |
| | III | Steel construction with fire resistance<br>Masonry with fire resistance<br>Reinforced concrete with fire resistance | Nonfuel contributing and stable to the degree of fire resistance |

Table reproduced from *Steel and Fire Safety — As Required in the National Building Code of Canada 1975*, Roger V. Hebert, Canadian Steel Construction Council.

## Type I Construction

In Type I or fire-resistive construction, the structural members are of noncombustible materials that have a specified fire resistance. Generally, bearing walls, columns, and beams are required to have a fire resistance of three or four hours. Floor construction is required to have a fire resistance of either two or three hours. Roof construction is either one or two hours. Partitions that separate occupancies or tenants may also be required to be fire resistant as specified by the local code. The building codes provide some limited exceptions to the very restrictive requirements of this type of construction and usually permit the limited use of combustible materials in the building.

Because most of the actual structure in a fire-resistant building is noncombustible, the contents contribute most of the fuel for a fire (Figure 3.1). A fire in a vacant fire-resistive building will be of limited magnitude. From the fire investigator's point of view, a fire-resistive building is designed to withstand a fire in one or more of its compartments with little or no structural damage. Potential avenues for fire and smoke travel include:

- Penetrations in floor/ceiling assemblies for building services

- Spaces or voids between floors in exterior wall system and floor/ceiling assemblies

Large fires in Type I buildings have been known to damage or destroy protected structural members. In these situations, the fire investigator should assess the damage in that fire area and seek expert advise if there is any question as to the integrity of the building or its components.

**Figure 3.1** Fires in Type I construction primarily involve the contents of the building. *Courtesy of Ed Prendergast.*

## Type II Construction

Type II construction is also known as noncombustible construction. Type II construction can be either protected or unprotected. In unprotected construction, the major components are noncombustible but have no fire resistance. The use of unprotected steel is the most common characteristic in unprotected, noncombustible construction (Figure 3.2).

Frequently, structural steel is provided with a degree of fire resistance that is less than that required for Type I construction. This is sometimes known as protected, noncombustible construction. Referring to Table 3.1, it can be seen that Type II construction is divided into three subclassifications. Two of the subclassifications, designated 222 and 111, have structural components with one- or two-hour fire-resistance ratings. The third subclassification, designated 000, has no fire-resistance requirement.

Materials other than steel can be used in Type II construction. A material such as concrete block can be used for the walls of a Type II building with steel beams used to support the roof. Glass and aluminum can be used, but their structural role is limited.

**Figure 3.2** Typical Type II construction. *Courtesy of Ed Prendergast*

Building codes allow the use of combustible material in Type II buildings for applications similar to those in Type I construction.

Type II construction is widely used for commercial, storage, and industrial properties. From the fire investigator's perspective, the most notable issue with this type of construction is the failure of steel framing members after relatively limited exposures to heat. The fire protection measures used on the structural members to achieve the various ratings will protect them for a period of time, but this type of construction will not withstand the effects of a fire as well as will Type I construction. Steel framing members regularly used in the construction of a Type II building lose their strength when heated above 1100°F (593°C). When heated, steel not only loses strength, it also expands slightly. Thermal expansion can cause beams to exert unexpected forces on bearing walls that can cause them to fail. These two physical properties of steel make Type II construction very susceptible to collapse as a result of fire exposure. Structural collapse can make the fire investigator's job much more difficult and dangerous because large volumes of debris may have to be moved, and portions of the structure may be unstable. Other issues the fire investigator must consider with Type II construction include the use of foamed plastics as insulation on the walls and roof deck and the presence of numerous open channels that can allow the fire and smoke to travel about the building.

## Type III Construction

Type III construction is commonly referred to as ordinary construction. Type III construction usually is constructed with exterior walls of masonry. From a technical standpoint, however, any noncombustible material with the required fire resistance could be used to construct the exterior walls. Interior structural members including walls, columns, beams, floors, and roofs are permitted to be partially or wholly combustible (Figure 3.3). Even though many of the structural components in Type III construction may be provided with some fire protection, the fire investigator should consider them as part of the available fuel in the event of a fire.

A fundamental fire concern with Type III construction is the combustible concealed spaces that are created between floor and ceiling joists and

between the studs in partition walls when they are covered with interior finish materials. These spaces provide combustible paths for the communication of fire through a building. Fire can enter these spaces when openings exist in the interior finish materials or when a fire is of sufficient magnitude to destroy the material.

## Type IV Construction

Type IV construction is commonly known as heavy timber construction. Like Type III construction, the exterior walls are normally of masonry construction, and the interior structural members are combustible. Two distinctions exist between Type III and Type IV construction. In Type IV construction, the beams, columns, floors, and roofs are made of solid or laminated wood with dimensions greater than in Type III construction (Figure 3.4).

**Figure 3.3** Interior structural members in Type III construction are wholly or partially combustible. *Courtesy of Ed Prendergast*

**Figure 3.4** Typical Type IV construction. *Courtesy of Ed Prendergast*

**Table 3.4**
**Recommended Nominal Dimensional**
**Requirements for BCMC Type IV (2HH) Construction**

| | Supporting Floors | Supporting Roofs |
|---|---|---|
| Columns | 8 in. x 8 in. | 6 in. x 8 in. |
| Beams and girders | 6 in. x 10 in. | 4 in. x 6 in. |
| Arches | 8 in. x 8 in. | 6 in. x 8 in., 6 in. x 6 in., 4 in. x 6 in. |
| Trusses | 8 in. x 8 in. | 4 in. x 6 in. |
| Floors | 3 in. T & G or 4 in. on edge w/1 in. flooring | |
| Roofs | | 2 in. T & G or 3 in. on edge or 1⅛ in. plywood |

SI units: 1 in. = 25.4 mm

*Reprinted with permission from the Fire Protection Handbook, 18th edition, copyright© 1998, National Fire Protection Association. All rights reserved.*

The minimum dimensions typically required by a building code for wood members in Type IV construction are shown in Table 3.4.

Heavy timber construction was used extensively in factories, mills, and warehouses in the nineteenth century. It is rarely used today in new construction, although many old buildings of this type remain in use. The primary fire hazard associated with Type IV construction is the massive amount of fuel presented by the large structural members in addition to the building contents.

## Type V Construction

In Type V construction, all major structural components are permitted to be of combustible material. Type V construction is also known as wood-frame construction. The primary method of construction in a Type V building consists of using a wood frame to provide the primary structural support. This is the common type of construction used for residential properties in North America and is one that the fire investigator will see regularly (Figure 3.5).

Table 3.1 shows that there are two subclassifications in Type V construction. One of the subclassifications requires one hour of fire resistance for the structural members; this is typically accomplished by covering the combustible frame members with plaster or fire-rated gypsum board.

A fundamental problem posed by Type V construction is the creation of more extensive combustible concealed voids and channels than is found in Type III construction. These concealed spaces provide serious potential for the extension of fire within a building. Because it is inherently combustible, a Type V building can become totally involved and completely destroyed in a fire. A heavily involved wood-frame building also poses a threat to adjacent buildings. Building codes impose restrictions on the maximum allowable heights and areas of wood-frame buildings. The codes also require spatial separations between wood-frame buildings.

**Figure 3.5** Typical wood-frame (Type V) construction. *Courtesy of Ed Prendergast*

As with any construction type, the fire investigator must be alert to damage to the structural members that make up the Type V building. Framing members may have been burned away, leaving only a floor or roof covering in place with little or no support. Major portions of the building may also be unsupported as a result of framing members being burned away. The fire investigator may find that fire and smoke are able to travel from floor to floor and room to room via non-fire-stopped spaces in the framing members.

 ## Other Types of Construction

### Mobile Home Construction

The typical mobile home is an assembly of four major components: the chassis and the floor, wall, and roof systems. Although they are constructed of steel, wood, plywood, aluminum, gypsum wallboard, and other materials, they are basically frame construction. About 70 percent of all mobile homes conform to a general design; single-section units (single-wide) are typically 14 feet (4.3 m) wide and between 55 and 70 feet (17 m and 21 m) in length. The remaining 30 percent consist of multiple sections (multiwide): two or more sections that when set on a permanent foundation can appear almost indistinguishable from houses of conventional construction. Sizes range from 24 to 28 feet (7.4 m to 8.5 m) wide, and from 60 feet to 70 feet (18 m to 21 m) long. A crawl space is an option, as is a basement. Usable living space ranges from 966 square feet (89.7 m²) for a typical single-wide to 1,440 square feet (133.7 m²) for a typical double-wide mobile home.

The roof construction of most newer mobile homes uses the shallow truss for structural support. Even though these trusses are designed to support a minimum load of 1,000 pounds (453.6 kg) per pair, when placed in a conventional supporting method, their integrity should always be monitored. This type of construction forms a void between the roof material and the interior ceiling allowing an avenue for fire extension. Thirty-gauge sheet metal is the most common material used for roofs, while low-density, Class C fire-resistive fiberboard is used as the interior ceiling material. These void spaces can allow for possible fire communication to other spaces in the structure.

In the last two decades, changes in the design and construction of manufactured homes (as mobile homes are now commonly called) have reduced fire hazards related to these units. The driving force for change began in 1976 when HUD released standards (3280.203–206) which specified designs that limit the spread of fire. The HUD standards require a smoke detector placed between bedrooms and other living spaces. Some of the additional requirements of the standards include:

- Foam plastic insulating materials cannot be used in walls or ceilings unless very strict requirements are met.

- Cooking areas above and behind the range must be constructed of materials that have limited combustibility (flame spread ratings no greater then 50). If cabinets are placed above the range, a metal hood is to be installed with 5/16 gypsum board or the equivalent (unless one-inch framing is used) between the hood and cabinet.

- Furnace and water heater compartments must have interior finishes that have a flame spread rating no greater than 25 and be made with limited combustible materials. Carpets may not be placed below furnaces or water heaters.

- Every newly manufactured home receives a dielectric test (hi-pot) at the factory. In this test, a voltage of at least 1000 volts is applied to each of the hot wires in-turn. Any substantial leakage current (through damaged insulation or an incorrect wiring connection) is cause for the system to fail the test.

The HUD standards have resulted in improved designs, the use of fire-resistive construction materials such as gypsum, installation of smoke detectors, and the inspection of each unit prior to shipment. Due to these improvements, today's manufactured homes are safer than older models. The unique construction of manufactured homes does, however, impact the behavior of the fire within them. The fire investigator should consider the construction standards as they investigate the fires in this type of unit. The investigator must be concerned with the features of manufactured homes, including the metal roof and walls, and their influence on the containment of heat and the development of the fire within a unit.

## Prefabricated Construction

Prefabricated or panelized construction defines a method of building construction where the walls, floors, and ceilings are manufactured complete with plumbing, electrical wiring, and all millwork. Once delivered to the site, the entire assembly is erected. Being lightweight in construction promotes a quick erection time. Though structurally sound, this form of manufactured construction is generally associated with rapid fire spread due to the number of common voids.

Lightweight beams and trusses are also susceptible to very rapid failure when exposed to fire. The fire investigator should take this into consideration when conducting an investigation. The amount of damage may not appear to be consistent to the amount the fire investigator would find with heavier framing members.

## Modular

A modular building is built in two or more sections at the factory. All utilities, such as plumbing, heating, and electrical systems, are installed as an integral part of the construction phase. In addition, all millwork, such as doors and windows, is installed by the manufacturer. Some modulars are designed to be stacked together to form multiple-story buildings. As in most manufactured buildings, lightweight components are used. The components are structurally sound as long as they are not attacked by fire. Breaching of any structural member in this type of construction can promote structural collapse.

Modular buildings are normally constructed to meet or exceed most of the model building codes. This, coupled with the need to construct the building to withstand transportation from the factory to the location where it will be erected, usually results in a building that is relatively strong. The fire investigator may also find unusual paths of fire travel that are due to the modular method of construction. These paths could include chases and shafts for heating, plumbing, and electrical services that would not be found in buildings that are site built.

## Geodesic Construction

A geodesic structure is defined as a dome or vault made of lightweight straight structural elements installed so as to form a tension load. The principle of geodesic construction is to reduce the weight at the tension points to make the structure economical to build. However, being domed or hemispherical in shape promotes a greater rate of fire spread and difficulty in laddering the structure. These factors can lead to a rapid degradation of structural stability resulting in structural collapse.

## Log Homes

Log homes are another style of premanufactured housing. These homes are constructed from solid logs ranging from 4 to 9 inches (101 mm to 228 mm) in diameter. Two sides and both ends of each log are machined to form a tight, structurally sound fit. Because of their mass and surface area, the walls of log homes are naturally well-insulated. Under fire conditions, this can result in higher-than-expected temperatures in the building and less ventilation from the exterior through the walls. Varnish and other preservatives applied to the interior surfaces of the logs could also result in unusual burn patterns after a fire.

 ## Building Materials

In the discussion of construction types, several different types of building materials were discussed. In modern practice, the materials most commonly used for the major structural components of buildings include:

- Wood
- Masonry
- Concrete
- Iron and Steel
- Aluminum
- Glass

Each of the materials identified above is discussed in more detail in the following sections.

Building materials have a variety of properties, such as strength, density, appearance, durability, thermal conductivity, and resistance to corrosion and insects, that determine their usefulness in various architectural applications. The fire investigator should understand the basic properties of these materials and the effect that exposure to heat and flame has on them. Some of the basic properties of materials the investigator should know include the following:

- Combustibility
- Thermal conductivity
- Rate of thermal expansion
- Effects of heat on the material

## Wood

Wood products are probably the most widely used construction materials the fire investigator will encounter. Wooden structural members are found in Type III, Type IV, and Type V construction. The primary difference in their uses is the dimension of the members. For Type III or ordinary construction, the minimum nominal thickness of any component is 2 inches (50 mm) with varying widths depending on the load being supported. For Type IV or heavy timber construction, beams and girders must have nominal dimensions of not less that 6 inches (150 mm) in width and 10 inches (250 mm) in depth. Type V construction requires significantly smaller dimensions for the materials used in construction.

As discussed in Chapter 2, "Fire Behavior," the mass and surface area of a fuel affect how rapidly it burns. Wood in its normal state is combustible. However, from a structural integrity perspective, a 6-×-10 inch (150 mm by 250 mm) beam will support its load for a significantly longer time than one with dimensions of $2 \times 6$ inches (50 mm by 150 mm).

Wood is not a good conductor of heat, and it does not expand significantly when heated. Thus, two primary properties of wood with which the fire investigator must be concerned are combustibility and the ability of the structural members to support the load of the building after they have been exposed to fire.

## Masonry

Masonry is one of the oldest and simplest of building techniques. The use of masonry dates back thousands of years. The fundamental construction technique consists of stacking the individual masonry units on top of one another and bonding them with mortar into a solid mass. Masonry is a durable building material, and masonry structures that have stood for centuries are not uncommon.

Masonry units can be made of several materials including:

- Brick
- Concrete block
- Stone
- Clay tile block
- Gypsum block

Masonry construction units are not combustible; they tend not to expand due to heat exposure. The primary problem with this form of construction is the deterioration of the mortar used to bond the units together. Exposure to water (including fire streams) can cause the mortar to deteriorate and result in collapse. Some masonry units, such as natural stone, are also susceptible to spalling when exposed to fire.

## Concrete

Concrete has many applications in building construction. It is used for pavement, foundations, columns, floors, walls, and the concrete masonry units discussed in the previous section. Its advantages are that it can be made from raw materials (which are usually locally available and are low in cost) and that like masonry, concrete does not burn, and it resists insects and the effects of contact with soil. It can be placed in forms to create a variety of architectural shapes.

Concrete is produced from portland cement, coarse and fine aggregates, and water. The aggregates used in concrete are inert mineral ingredients that reduce the amount of cement that otherwise would be needed. The course aggregates consist of gravel or stone, and the fine aggregate is sand. The cement combines with the water to form a paste that coats and bonds the pieces of fine and coarse aggregate. The aggregates make up a large percentage of the total volume of concrete.

Because concrete is weak in tension, it cannot be used alone where tensile forces occur in a structure. To resist the tensile forces, concrete is reinforced through the use of steel reinforcing bars placed within the concrete before it hardens. Steel has a high tensile strength and expands with changes in temperature at the same rate as concrete.

Concrete is a fundamentally fire-resistive material. It is noncombustible, and it has good insulating properties. Although the steel used in reinforced concrete is not fire resistive, the concrete that surrounds it acts as insulation to protect it from the heat of the fire. The overall fire resistance of the reinforced concrete depends on the depth of cover of the concrete over the steel and the quality of the concrete.

One of the most significant effects fire and heat have on concrete is spalling. *Spalling* of the concrete is caused primarily by the expansion of the excess moisture either when it is heated or when it freezes. The expansion of the water creates tensile forces within the concrete. Because concrete has little resistance to tension, small pieces of the concrete break off. This spalling weakens the concrete structurally and exposes the reinforcing steel to the fire. As a general observation, concrete that is of poor quality structurally will also perform poorly when exposed to the fire.

## Iron and Steel

Steel is the strongest of the structural materials. Steel is a material that is nonrotting, resistant to aging, and dimensionally stable. The complex industrial process by which it is produced is subject to tight control, resulting in a product with generally consistent quality. Steel is a relatively expensive material, but its strength and the variety of forms in which it is produced permit it to be used in smaller quantities than other materials. Steel is used in the construction of buildings for applications varying from heavy beams and columns to door frames and nails.

For the fire investigator, the deterioration of the strength of steel at elevated temperatures is its most significant characteristic. The fires normally encountered by the fire service do not create temperatures hot enough to melt steel. However, they are hot enough to greatly weaken steel. Because temperatures in excess of 1,200°F (650°C) (the temperature at which steel loses most of its strength) are regularly encountered in fires, failure of unprotected steel to a greater or lesser degree can be anticipated. The loss of strength because of increased temperature is not a sudden occurrence taking place at one temperature. Rather the steel loses its strength gradually as its temperature increases.

The quickness at which unprotected steel fails when exposed to the fire depends on several factors including:

- Mass of the steel members
- Intensity of the exposing fire
- Load supported by the steel
- Type of structural connections used to join the steel members

Other properties of steel that can be a factor in a fire include:

- *Thermal expansion.* Heat will cause structural elements to expand significantly which can cause the collapse of walls.
- *High conductivity.* Steel is a very good conductor of heat. Elevated temperatures at one end of a beam or column will be transmitted through the member. This could result in the ignition of combustibles some distance from the point of initial exposure.

Cast iron was used as structural framing in buildings before the turn of the century. The fire investigator may still find cast iron columns in older buildings. A few structures were built with complete cast iron fronts. However, in modern practice, cast iron has been completely replaced by steel. Because cast iron is a brittle material, it tends to fail by fracturing rather than by yielding as in the case of steel.

## Aluminum

Aluminum is used in conventional residential and commercial construction and in manufactured homes. In conventional residential construction, aluminum is used primarily in window and door frames and as roof panels and siding. In commercial construction, aluminum is often used in window and door frames but is used most extensively in curtain walls on the exterior of high-rise buildings. Aluminum is also used extensively in the construction of mobile homes. Because these structures are designed to be transported from place to place, aluminum's high strength-to-weight ratio makes it a desirable construction material. It is used in the construction of window and door frames as in conventional construction, but it is also used as roof covering, exterior wall covering, and in some older units, for electrical wiring.

# Glass

Glass is present in most buildings. Its obvious use is for windows, skylights, storefronts, and other applications where the transmission of light is desirable. The architectural applications of glass extend to gothic church windows, partition walls, and the exterior curtain walls of buildings.

As is the case with other building materials, several different types of glass are produced. The most commonly encountered glass types include ordinary, single-strength annealed; tempered; heat-strengthened; laminated; and glass block.

- *Annealed*. Annealed glass is produced by slowly cooling the hot glass during its production. The slow cooling permits the release of thermal stresses that would form if the glass were cooled rapidly.

- *Tempered*. Tempered glass is about four times stronger than annealed glass. It is used in windows that might be subject to high wind forces and in exterior doors that people might walk into. Tempered glass is produced by cooling the exterior surfaces with air while allowing the inner core to cool more slowly. This process results in compressive stresses in the edges of the glass that give it greater strength. When tempered glass is broken, the internal stresses are released, producing small granules of glass rather than large sharp-edged pieces or shards.

- *Heat-strengthened*. Heat-strengthened glass costs less than tempered glass, but it is not as strong. Compressive stresses are also induced in heat-strengthened glass, but the stresses are not as great as those in tempered glass.

- *Laminated*. Laminated glass consists of two layers of glass with a transparent layer of vinyl bonded into the center. When the glass breaks, the inner core of vinyl holds the broken pieces of glass in place. Laminated glass is a good barrier to sound and, therefore, can be used to reduce noise transmission. Firefighters may encounter laminated glass in security windows used for drive-in bank tellers and similar applications.

- *Glass block*. Glass blocks are produced as solid or hollow units with different surface patterns that create varied light patterns. The blocks typically are available in sizes ranging from 6 x 6 inches (150 mm by 150 mm) to 8 x 8 inches (200 mm by 200 mm) and 3 or 4 inches (75 mm or 100 mm in thickness). Glass block is not load bearing. Any masonry or other structural load above a glass block assembly must be supported independently of the glass block. The individual glass blocks are assembled into panels with either mortar or a silicone sealant between individual blocks. The glass block panels are secured to the surrounding wall using steel channels or specially designed anchors.

Glass is noncombustible but is not fire resistive. When heated, internal thermal stresses cause glass to shatter and fall out of its frame. However, there are two types of glass that are suitable where fire resistance is required. These types are wired glass and the more recently developed fire-rated glass.

## Wired Glass

Wired glass is made by rolling a mesh of wires into a sheet of hot glass. When the glass breaks, the wires hold the glass in place permitting it to act as a barrier to the fire. Wired glass is used in both interior and exterior applications. It is used in fire doors, in windows adjacent to fire escapes, and in corridor separations. It is also used to protect against exterior exposures.

## Fire-Rated Glass

Within the last ten years, advances in technology have allowed glass manufacturers to produce products that can achieve fire test ratings of one hour or more. These products do not use wire to keep the material in place but are a combination of plastics and glass. Because these products are relatively new to the building industry, fire investigators may not see them on a regular basis. The investigator should, however, know that they are in use and understand that they will react very differently to fire exposure.

## Crazing

Fire investigators often will encounter cracked or broken glass at a fire scene. It is important to understand the mechanisms that cause breakage and to incorporate that information into the investigation findings.

Glass will crack when temperature variations of over 140°F (60°C) occur between exposed surfaces and insulated portions. These cracks typically appear to be long, smooth, and wavy, and radiate across the pane. As the cracks form, they can cause the pane to fail, allowing broken portions to fall from the frame.

The *crazing* of glass is the formation of patterns of short cracks throughout the pane. Crazing can occur when water is suddenly applied to one side of a hot pane of glass. Fire fighting operations thus could result in crazing should an attack stream be played on the surface of a hot window.

Glass will also break as a result of excessive pressure on its surface. Research shows that the pressures necessary for breakage to occur are approximately 0.3 psi to 1.0 psi (2.07 kPa to 6.90 kPa). The pressures developed by a fire in a compartment are significantly lower that this and would not be sufficient to cause the glass to break. However, pressures from explosions, including backdrafts, can be high enough to break glass.

## Plastic Construction Materials

The term *plastic* encompasses a large number of synthetic organic materials of high molecular weight that can be formed by pressure, heat, extrusion, and other methods. There are 20 to 30 major groups of plastics. In addition, variations can be produced within the major groups by varying the chemistry of individual materials. The large variety of plastics available permits their use in many different applications. In building construction, plastics are used for such components as:

- Siding
- Floor covering
- Insulation
- Tub and shower enclosures
- Vapor barriers
- Pipe and pipe fittings
- Lighting fixtures
- Skylights and roof domes
- Sprinkler piping

The use of plastics in building construction increases the fire hazard to the extent that it increases the amount of fuel in a building or the toxicity of the products of combustion. Therefore, the flammability of plastics is of interest to the investigator. As with their other properties, the flammability of plastics varies widely. Some plastics, such as cellulose nitrate, burn so rapidly that they constitute a unique fire hazard. Other plastics may burn slowly and stop burning when the ignition source is removed. Fire retardants can be added to some plastics to reduce their flammability and ignition sensitivity. However, even plastics with low flammability are subject to pyrolization and may produce toxic gases at temperatures above 500°F (260°C).

Plastics materials frequently exhibit burning properties different from other materials. For example, nylon usually melts and drips when it burns. Foam plastics burn more intensely when tested on a large scale than when tested in small samples. Some plastics generate enormous quantities of heavy smoke. The products of combustion of some plastics are more toxic than nonplastic materials. The combustion of vinyl chloride, for example, produces hydrogen chloride — the gaseous form of hydrochloric acid. Hydrogen chloride is corrosive as well as toxic and increases the damage done to sensitive electrical equipment.

 **Building Components**

Up to now, this discussion has focused on the structural framing and types of materials used to form that frame. Should fire attack the frame of any building, serious damage can result as well as structural collapse. From the fire investigator's point of view, there are several additional components of a structure that he should understand. Those components include:

- Roof systems
- Foundations
- Interior finish
- Compartmentation

### Roof Systems

The basic purpose of a roof is to protect the inside of a building from exposure to snow, wind, rain, etc. However, the roof also provides for a controllable interior environment, enhances the architectural style of a building, and contributes to the functional purpose of the building. Roofs are frequently described by their style or shape. Typical roof shapes are shown in Figure 3.6.

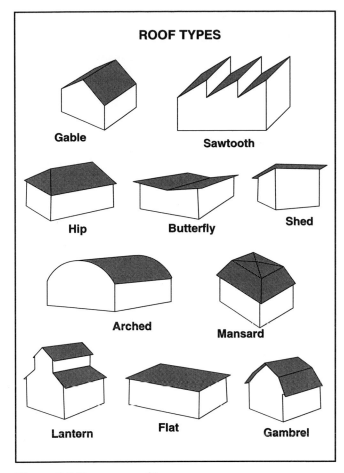

**Figure 3.6** Common roof types.

The fire investigator should use caution when examining a roof because the structural stability of the system may have been compromised by the fire. Roof systems can collapse in large sections, making the examination of the interior of a building very difficult. In these cases, the use of heavy construction equipment may be the only safe method to remove the debris.

Also of interest to the fire investigator are roof systems built from products that will melt and flow when exposed to the heat of a fire. Many of these products are made from petrochemicals and will burn when exposed to fire. The flowing material can result in burning that is remote from the area of origin and give indicators that are similar to those of an accelerant.

## Roof Supporting Systems

Roof systems are usually designed to withstand a live load of 20 to 25 pounds per square foot (.96 kPa to 1.2 kPa) depending on the local code. A *live load* is any load placed on the building that is not part of the structure. The load that results from structural

components is known as *dead load*. Roofs must also be designed to withstand wind forces. The local building code may require that a roof withstand a specified internal pressure to resist lifting forces that the wind may exert. In general, roof systems are not designed to carry the live loads that floors carry

*Pitched Roofs.* A *truss* is a framed structural unit made of a group of triangles in one plane. If loads are applied at the points of intersection of the truss members, only compressive or tensile (nonbending) forces will result in the members. The basic triangles of which trusses are composed can be arranged in a large variety of styles. Standard truss shapes are available to span distances of 22 feet to 70 feet (6.7 m to 21.3 m). Typical pitched roof types are gable, shed, hip, butterfly, sawtooth, gambrel, and mansard.

*Flat Roofs.* Flat roofs are usually supported by roof joists. The technique is structurally similar to that used for floors. Roof joists can be wood beams or steel bars. The roof deck sheathing is attached to the joists. In addition to wood, roof decks can be made of precast gypsum, concrete planks, wood-fiber cement planks, and light-gauge steel-ribbed panels.

*Arched and Domed Roofs.* Arches are suitable for supporting roofs with large, clear spaces, such as exhibition halls and field houses. Arches were originally constructed of masonry, but modern arches can be made of wood, concrete, or steel. When the area to be enclosed by the roof is circular, a dome roof can be used. A dome roof produces structural forces similar to those of an arch. That is, horizontal thrusts exist at the base and a compressive force exists at the top. To support the outward thrust at the base of the dome and keep it from spreading, a structural member known as a tension ring is provided. At the top of the dome the forces are inward and a compression ring is used (Figure 3.7). Similar to arches, domed roofs can be made of concrete, steel, or laminated wood

## Roof Coverings

A wide variety of roof coverings are in use. Roof covering materials include sheet metal (galvanized steel, copper, aluminum, and lead), clay tile, slate, plastic coatings, and felt and asphalt in shingle and roll forms.

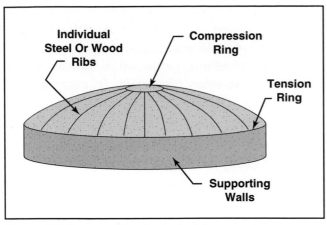

**Figure 3.7** Components of a typical dome roof.

Roofing is frequently of a built-up form. This consists of layers of roofing felt applied to a roof deck with intervening layers of roofing cement. The layers of felt are topped with a layer of tar and gravel.

Wood shingles are also used as a roof covering in many parts of the country. Many times in drier climates wood shingles are applied to the structural members of the roof using no intermediate sheathing. Wood shingles are a significant hazard should flying brands or other burning materials fall on them and cause them to ignite. This problem is well documented through fires from colonial times to the present. Attempts to regulate their use have almost always met with opposition from the wood industry and home owners concerned with aesthetics.

## Foundations

The function of a foundation is to transfer the structural load of a building to the ground. Ordinarily, the structural problems of foundations are of little interest to the investigator. However, the failure of the foundation can create or aggravate structural problems within the building supported by the foundation. Supports that shift or settle alter the forces on the structural members in the upper part of a building. In severe cases, the frame of a building may be distorted, floors may slope, walls and glass may crack, and doors and windows may not work properly. In some cases, automatic sprinkler pipe can be damaged. These altered load patterns can also hasten structural collapse under fire conditions and may need to be evaluated by an engineer prior to proceeding with an investigation.

Another area of interest to the fire investigator is damage to the foundation that results from an explosion. The information presented in this section describes the construction of the foundation; evaluating damage is discussed later in the manual. The type of foundation a building requires depends on the type of building and the soil conditions at the site. For example, a small garage or shed requires only a simple foundation; a high-rise building requires a foundation that extends 100 feet (30.5 m) or more.

The foundation is designed to support the dead load of a building and the live load of its contents. In addition to these loads, a foundation may need to be designed to resist other forces such as the following:

- Wind loads that may apply lateral or uplifting forces to a building
- Soil pressure
- Uplifting forces from underground water
- Thrusts resulting from the support of arches, domes, or vaults
- Seismic forces

Foundations can be divided into two types: shallow and deep. A *shallow foundation* transfers the weight of the building to the soil at the base of the building. A shallow foundation can be used where the load-bearing ability of the soil directly under the building is adequate to support the building. *Deep foundations* penetrate the layers of soil directly under a building to reach soil at a greater depth that can support the weight of a building.

The foundation of a building with a basement must have walls enclosing the basement. Concrete is the material most commonly used for foundation walls. Concrete is durable and is resistant to moisture and insects. Many concrete foundation walls develop visible cracks for varying reasons. They usually do not significantly affect the ability of the wall to support or distribute the load which it is carrying. However, when inspecting a structure for stability, any change in size or extension of cracks or fissures should be given close attention. Any vertical or horizontal misalignment along the length of a crack in a foundation wall indicates a movement or shift in the structure, which may mean a change in the way loads are being transmitted from structural members to the foundation.

Foundation walls may be constructed of concrete, stone, brick, or concrete block. Stone and brick foundation walls are found usually in older buildings. A distinguishing aspect of stone foundations is that they were often constructed without using any bonding mortar or cement. The stones were carefully quarried and meticulously assembled to form an amazingly tight-fitting and strong foundation.

## Interior Finish

An important consideration for the fire investigator is the combustibility of the materials used for interior finish. Throughout history, combustible interior finish materials have been the cause of numerous fires involving a high number of lives lost (Table 3.5).

The NFPA's *Fire Protection Handbook*, 18th edition, identifies the following four ways in which a building's interior finish relates to a fire:

- It affects the rate of fire buildup to a flashover condition.

- It contributes to fire extension by flame spreading over its surface.

- It adds to the intensity of a fire by contributing additional fuel.

- It produces smoke and toxic gases that can contribute to life hazard and property damage.

As we have already discussed, fire behavior is highly dynamic and influenced by several variables. Therefore, in terms of practical applications, it is difficult to obtain precise measurements of the surface burning characteristics of materials. The speed of flame spread over an interior finish material will be influenced by such factors as the chemical composition of the material used as a backing and the geometry of the space in which it is installed.

A building code specifies what classes of material can be used for interior finish in various occupancies or portions of a building. For the most part, the most stringent requirements for interior finish are in spaces used for egress from the building. The classes of materials used for interior finish are based on flame spread and smoke developed during testing. Table 3.6 shows the building code classifications for wall and ceiling coverings.

Interior finish materials with better ratings (Class A), such as painted gypsum board, would be expected to perform better than those with lower or no rating, such as thin wood paneling or nonrated carpet applied to a wall surface. Class A and B materials generally contribute little or no fuel to the development of a fire (Figure 3.8).

| Table 3.6 | |
|---|---|
| Class A | Flame spread 0-25, Smoke developed 0-450 |
| Class B | Flame spread 26-75, Smoke developed 0-450 |
| Class C | Flame spread 76-200, Smoke developed 0-450 |

| Table 3.5 Major Fires Where Interior Finish Was a Factor in the Large Loss of Life | | |
|---|---|---|
| Coconut Grove | November 28, 1942 | 492 dead |
| The LaSalle Hotel | June 5, 1946 | 61 dead |
| Canfield Hotel | June 9, 1946 | 19 dead |
| Winecoff Hotel | December 7, 1946 | 119 dead |
| Our Lady of the Angels School | December 1, 1958 | 93 dead |
| Hartford Hospital | December 8, 1961 | 16 dead |
| Beverly Hills Supper Club | May 28, 1977 | 165 dead |
| Dupont Plaza Hotel and Casino | December 31, 1986 | 98 dead |

**Figure 3.8** Class A materials hold up well in fires.

## Compartmentation

When an area of a building is subdivided with numerous fire-resistive partitions, walls, or floors, it is said to be compartmentalized. The fire-resistive partitions tend to contain fire and block horizontal spread. Fire rated floor/ceiling assemblies provide protection from the vertical spread of fire between floors of the building. The purpose of compartmentation is to try to localize the fire to its place of origin. This is done by creating a barrier to prevent the spread of fire and fire by-products. Examples include the use of rated partitions, fire doors, rated floor slabs, and ceiling assemblies. Fire-resistive compartmentation, while beneficial, provides only passive fire protection. That is, it may block or retard a fire, but it cannot extinguish the fire. For extinguishment, either hose streams or automatic sprinklers are necessary.

Building codes typically require fire-resistive partitions or separations in locations such as corridor walls, stairway and elevator-shaft enclosures, occupancy separations, and between building floors. Fire-resistive separations can be constructed from a wide variety of materials, including wire lath and plaster, gypsum wallboard, concrete block, and combinations of materials. The degree of fire resistance provided depends on the material used and its thickness.

The difference between a fire-resistive partition wall and a fire wall is a matter of degree. A fire wall is designed to withstand a severe fire exposure and to act as an absolute barrier against the spread of fire. A fire wall usually has fire resistance of four hours, with all openings protected by automatically closing fire doors. A fire-resistive partition will have a fire resistance less than that of a fire wall. The purpose of fire walls is to subdivide a building into areas so that a fire in one area will be limited to that area and not destroy the entire building. Fire walls are always masonry. They are customarily designed to be self-supporting so that structural collapse on either side of the fire wall will not damage it. Because fire walls are intended to be absolute barriers against fire, no combustible construction can be permitted to penetrate the wall. Where combustible floor and roof beams abut a fire wall, they cannot pass completely through the wall. In addition, the fire wall must extend beyond combustible walls and roofs to prevent the radiant heat of flames from igniting adjacent surfaces. This is accomplished by topping the fire wall with a parapet. The parapet height above the combustible roof will be determined by the building code.

Depending on the type of construction, the building code may require that the separation between floors be rated (see Table 3.1). In all buildings, the floor system has a structural purpose. The floor is designed to support the live loads of the building. Floors may be constructed of wood or concrete. The supporting members (joists, beams, girders, or trusses) can be of wood, steel, or concrete depending on the construction type, load bearing capacity, and required fire rating (Figures 3.9 a and b).

The ceiling assembly has no structural role in the building. It is designed to support only its own weight and the weight of any fixtures attached to it. In rated construction, the ceiling system does provide a level of defense from fire to the floor assembly (deck and supporting structural members) above. Floor ceiling assemblies are rated and tested as a unit for fire resistance. Examples of specific systems and their ratings can be found in the *Fire Resistance Directory,* published by Underwriters Laboratories Inc.

The effectiveness of the floor ceiling assembly depends as much on adherence to proper installation requirements as on the materials used. Deviations from the installation requirements can result in the failure of the system to perform as designed.

The fire investigator should also consider the concealed space between the ceiling and floor system as a void space capable of becoming a path of travel for fire and its products of combustion. Once

**Figure 3.9a** Typical wooden I beams used as floor joists.
*Courtesy of Ed Prendergast*

**Figure 3.9b** Open web trusses used as floor joists.
*Courtesy of Ed Prendergast*

the fire enters this space, it can travel to remote locations on the floor of origin and extend vertically through openings in the floor.

Fire-rated construction intended to prevent the spread of fire is only effective if it is properly built and maintained. Penetrations in the assemblies can result in the failure of the assembly to perform as intended. Penetrations such as doors, windows, and pipe or cable passages can be designed into an assembly. These intended penetrations must also be rated and the assemblies maintained. If these passages are modified or overridden, such as wedging open a fire door in a rated wall, the protection is nonexistent. One of the most dangerous practices is the penetration of an assembly without proper provision to maintain the intended rating. This occurs when a hole is poked through a wall to run wires or pipe, or larger openings are made to allow passage. These conditions can result in the spread of the fire well beyond the compartment of origin. The fire investigator will address these issues when examining a structure where the fire has traveled beyond the room of origin.

## ◆ Conclusion

Building construction is such a broad category that it requires a certain amount of detail to cover it sufficiently. This brief overview has focused on some key elements that an investigator will need to build upon. Some of the main tenets that have been addressed include: types of construction, interior finishes, roof systems, building materials, foundations, and compartmentation.

The material presented here just scratches the surface. For greater detail, the reader should conduct self-study in the area of building construction. This can be in the form of texts such as IFSTA's **Building Construction Related to the Fire Service**, 2nd edition. An important thing to keep in mind is that construction techniques are ever-changing and that the investigator should try to keep abreast of new materials and building techniques.

# Chapter 4

# Building Services

Building services and the methods used for their installation may have a direct involvement in the ignition of fires in buildings. These services and their installation may provide the heat energy necessary for ignition, or they may serve as a source of fuel for the fire. The installation and location of these services can also affect how a fire and the products of combustion are transported in the building. The penetration of floors, walls, and ceilings by system wiring and/or piping often contributes to the spread of fires in buildings.

While not every service or system will be found in every building, these materials are presented as a background for the fire investigator. This chapter provides the fire investigator with an overview of each of the five major categories of building systems and explains how the installation, operation, or maintenance of the systems or services can affect a fire.

NFPA's *Fire Protection Handbook*, 18th edition, classifies building systems using five major categories:

- Energy production
- Environmental systems
- Piped services
- Electrical/electronic systems
- Transportation and conveyance

It is important for the fire investigator to remember that all these services or systems make up the total building "system." Many systems within today's buildings are interconnected and rely on input from other systems for proper operation. The fire investigator will have to determine how the systems in the building being examined may have been interconnected and what, if any, impact that interconnection had on the fire. The investigator must recognize the potential impact of any modifications made to a system after it is installed. Modifications, such as extending the system to cover building additions, may increase the likelihood of the system being involved in the ignition of a fire or the transport of products of combustion.

 **Energy Production**

Energy production systems include those that produce heat and cooling and provide electricity to buildings. The fire investigator can expect to find the following types of systems

- Electric generators
- Electrical substations
- Water heaters (Figure 4.1)

**Figure 4.1** Water heaters are examples of energy production systems.

- Boilers
- Air heaters
- Equipment used for refrigeration and cooling

Energy production systems take the utilities that enter a building and transform them into a usable state or distribute them for use in the remainder of the building. Examples of these systems include electrical substations and steam-reducing stations. Utilities, such as electricity, gas, or oil, that come into a building may also be converted by systems from one form of energy to another. An example of this is a boiler that uses electricity, gas, or oil to produce steam. Another example would be chillers that use electricity or some other energy source to produce chilled water for cooling. These systems are normally found in mechanical spaces that can be located almost anywhere within a building.

By their nature, energy production systems can provide either the heat energy or fuel necessary for a fire. Where heat is generated using the combustion process, it can be a source of ignition if the equipment is not properly operated and maintained, if good housekeeping practices are not followed near the devices, or if any combustible surfaces or materials are in the vicinity. If the fuel leaks from the device or associated piping that delivers it to the system, it can be ignited in a location not intended to contain the heat and fire.

Another potential source of fuel involving energy production systems are cooling towers. These units are commonly constructed of combustible materials. While in use, they are usually flooded and are unlikely to ignite; however, when they are not being used, they can become a significant fuel package on the roof of a building.

Another element of many energy production systems is the chimney or "stacks" used to transport products of combustion through the building to the outside atmosphere. Chimneys are often run in shafts that penetrate building floors and must be properly insulated to prevent the buildup of excessive heat. Both the chimney and the shaft it runs in must be constructed and maintained to minimize the leakage of hot and potentially toxic gases from the combustion process.

 ## Environmental Systems

The primary purpose of environmental systems in a building is to provide conditioned air to the building spaces for the comfort and safety of its occupants. These systems may range from small air-conditioning units that move relatively small volumes of air to large fans designed to move thousands of cubic feet of air per minute (Figure 4.2).

These systems may also be designed to remove odors and particulate materials from the air as well as to provide temperature control. Environmental systems may use "make-up air" from the outside or filter and recirculate the air within the building. These systems should be properly designed, installed, and maintained to minimize the transport of products of combustion throughout the building.

Environmental systems often use ductwork to move air around a building. The ductwork, as with other systems that service multiple spaces in a building, will be run in shafts and other concealed spaces in the structure. In order to prevent smoke or fire from traveling throughout a building via the ductwork, building and fire codes may require the installation of smoke and/or fire dampers within the ducts. Some air-handling systems are also interconnected to the fire alarm system and shut down automatically when a fire is detected. While the transport of air through ductwork presents little potential for developing sufficient heat to cause ignition, environmental systems may generate heat due to the fuels used for heating. Mechanical heating may be produced by the belts used to operate

**Figure 4.2** Air-conditioning is one type of environmental system. *Courtesy of Ed Prendergast.*

large fan units. In residential occupancies, combustible joist spaces may be used as the cold air return for the system, providing avenues for fire spread.

Environmental systems can provide a variety of potential fuel sources for ignition. The ductwork, duct coverings, filters, tape, and insulation used in the systems may be combustible. While the flame spread and smoke production of these materials are regulated by codes, there is a potential for the materials to become involved in fire. Other materials found in environmental systems are dust and other debris that collect within ductwork and on filters. Should these materials ignite and be of sufficient amount and continuity, there is a potential for fire spread over a wide area of a building.

##  Piped Services

Piping systems are used to transport a variety of gases and liquids within a building. Common building systems that use pipes include:

- Water
- Fire protection
- Heating (hot water or steam)
- Fuel gas
- Compressed air
- Medical gases
- Waste water/sanitary sewer

If the pipe system is carrying material that is ignitable, such as fuel gas, there is a potential for the

**Figure 4.3** Penetrations for piping can provide a means of fire spread. *Courtesy of Ed Prendergast.*

product to ignite should it leak from the system. Residential and light industrial fuel gas systems typically consist of natural gas (methane) or liquefied petroleum gas (butane or propane). The fuel gas system provides a combustible mixture to the intended energy release point (heat-producing appliance). The fuel is contained in piping, and the total system consists of all the equipment and pipe from the distribution point (tank or meter set) through the appliance. The installation and design of these systems are regulated by codes such as ANSI Z223.1-1999 *National Fuel Gas Code* (NFPA 54).

In fuel gas systems, the operating pressure (expressed in inches [kPa] of water column is regulated before entering a structure. Pipe may be metal or plastic (or a combination of both), and failures within the piping typically occur at joints or appliance connections. System failures may likewise occur within the appliance or pressure regulator, resulting in an accidental release of fuel and a potential of ignition at any heat- or energy-producing device. These failures may be a result of human (accidental or intentional), environmental (corrosion), or natural (earthquake, etc.) intervention.

Other common fire-related problems with piping systems are the chases and openings needed for the pipes to pass through walls and floors. If these penetrations are not properly fire stopped or otherwise protected, they can allow fire spread around them (Figure 4.3). Another potential with the growing use of plastic pipe is that of the fire actually consuming the pipe and passing through a barrier. Heat can also be transferred by conduction in metal pipe and cause the ignition of adjacent combustible materials some distance from the fire source.

##  Electrical/Electronic Systems

The electrical system provides electric current to power lights, machinery, equipment, and many of the other systems within the building. In recent years, there has also been a rapid increase in the amount of wire installed in buildings for telecommunications and computer networking.

The electrical system in a building consists of all equipment and wiring necessary to distribute power from the distribution point (part of the energy production system) to the remaining spaces. The design and installation of electrical systems are well regulated by codes such as NFPA 70, *National Electrical Code®*.

Wiring must be protected from physical damage by using conduit, shafts, or some other approved method. Where wire is run in spaces that could present a life safety risk should it burn, codes require the use of materials with a low smoke-production rating. Electrical systems have the potential to provide the heat necessary for ignition of combustibles as a result of arcing or overheating. This potential is discussed in more detail in Chapter 5, "Basic Electricity."

Telecommunications cable and equipment installed in buildings present a number of fire-related problems, including:

- Installation is often completed by unlicensed individuals with little knowledge of fire safety or other building systems.

- The volume of wire necessary in a building may represent a significant fire load.

- Cables are often run unprotected or are not listed for use in concealed spaces.

- Fire and smoke barriers are frequently breached to accomplish the installation of the cable.

- Associated equipment and workstations can be significant fuel packages.

- Loss of the communications system can result in significant business interruption losses.

## ◆ Transportation and Conveyance

These systems move occupants, materials, and components of other systems both vertically and horizontally in buildings. These systems include elevators, escalators, conveyors, trash and laundry chutes, and other materials-handling systems. Shafts for ductwork, piping, and wiring are also included in this category.

The transportation and conveyance systems provided in a building almost always require the penetration of floors and walls (Figures 4.4 a and b). The penetration of floors and walls by building systems can provide a path through which fire and

**Figure 4.4a** Elevator shafts often penetrate many floors in a building. *Courtesy of Ed Prendergast.*

**Figure 4.4b** Penetrations for escalators must be protected by automatic sprinklers. *Courtesy of Ed Prendergast.*

smoke can travel if not properly protected at the time of installation with automatic sprinklers or with fire dampers controlled by fusible links or smoke detectors. After installation, these penetrations must continue to be maintained to ensure that they do not provide a path of travel for smoke and fire. Building and fire codes contain provisions for the fire resistance of shafts that go from floor to floor. These requirements establish a level of protection to the remainder of the building should smoke or fire enter the shaft. For horizontal penetrations of building compartmentation or penetrations of floors with no shaft, the codes may require that the openings be fire-stopped or otherwise protected. Fire-stopping is intended to prevent smoke and fire from traveling through the openings. Other methods of protection include the use of water spray sprinklers on both sides of a horizontal opening and the use of doors or shutters that close on exposure to fire products.

Elevators are an essential system in many multilevel buildings, providing a means of transportation for both people and materials. They, however, present serious issues in the event of a fire due to the large shafts in which they operate. There is also the potential for occupants to be trapped as a result of the fire or by a loss of power. Because of recent legislation related to access to public buildings by individuals with physical disabilities, research is currently underway to make elevators available for evacuating these individuals in the event of a fire. Recent revisions to the NFPA *Life Safety Code*® provide requirements for elevators that are to be used as part of the egress system for a building. These requirements include provisions for a smokeproof shaft for the equipment and automatic sprinkler protection throughout the building.

Heat sufficient to ignite combustibles often comes from the electrical components of the system. There may be motors at various locations and in mechanical spaces that support the systems. As for fuel, the equipment may be combustible, there will be quantities of lubricants used for elevators and escalators, and there is the potential for the accumulation of combustible materials at the base of shafts provided as part of the system. In horizontal systems, such as conveyors, the penetration of fire walls can provide a means for fire and smoke to travel to other portions of a building. This potential can be increased if there are flammable or combustible materials transported on or stored near the conveyor.

## ◆ One- and Two-Family Dwellings

The services found in most one- and two-family dwellings may not be as complex as those found in larger buildings, but they can still affect a fire and the subsequent investigation.

Most homes will have several services installed to make them more habitable. These services can be classified as previously discussed. Figure 4.5

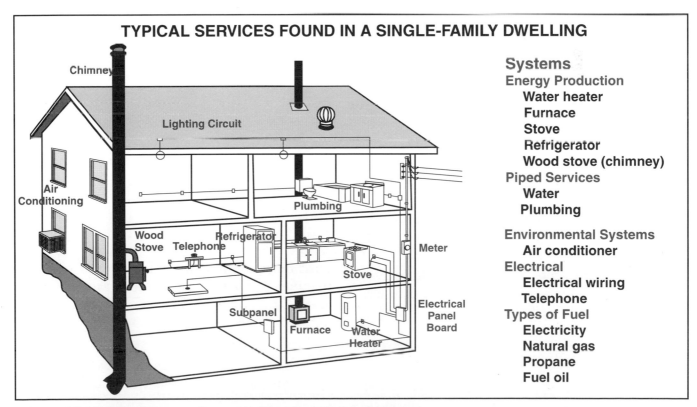

**Figure 4.5** Typical dwellings contain several different services and systems.

shows a typical dwelling with examples of the services and systems found in it. While they are on a smaller scale, the issues with building services are the same. The fire investigator must evaluate the impact the system or service had on a fire in a dwelling.

Questions the fire investigator should address regarding building services in fires involving residential properties include the following:

- Did penetrations for piping or wiring allow fire extension or smoke travel?

- Was the electrical wiring involved in the ignition of combustibles?

- Were the clearances from the furnace and wood stove according to code?

- What was the housekeeping like in the area of the heat-producing devices?

- Were the clearances provided for the furnace exhaust stack and the chimney for the wood stove proper to prevent ignition of structural components?

Because heating devices and their related chimneys are such a common issue faced by fire investigators in residential properties, they should have a very good understanding of the codes that regulate them. NFPA 211, *Standard for Chimneys, Fireplaces, Vents, and Solid Fuel-Burning Appliances*, provides the requirements for solid-fueled units. Table 4.1 provides information regarding proper clearances from combustible materials, and Table 4.2 provides information on the reduction of appliance clearances. Other applicable standards include:

- NFPA 31, *Standard for the Installation of Oil-Burning Equipment*

- NFPA 54, *National Fuel Gas Code®*

- NFPA 58, *Liquefied Petroleum Gas Code*

## Table 4.1
## Standard Clearances for Solid Fuel-Burning Appliances

| Kind of Appliance | Above Top of Casing or Appliance; Above Top and Sides of Furnace Plenum of Bonnet (in.) | (mm) | From Front (in.) | (mm) | From Back[3] (in.) | (mm) | From Sides[3] (in.) | (mm) |
|---|---|---|---|---|---|---|---|---|
| *Residential Appliances* Steam Boilers-15 psi (103 kPa) Water Boilers-250°F (121°C) max. Water Boilers-200°F (93°C) max. All water walled or jacketed | 6 | 152 | 48 | 1219 | 6[2] | 152[2] | 6[2] | 152[2] |
| *Furnaces* Gravity and forced air[4] | 18 | 457 | 48 | 1219 | 18 | 457 | 18 | 457 |
| *Room Heaters, Fireplace Stoves, Combinations* | 36 | 914 | 36 | 914 | 36 | 914 | 36 | 914 |
| *Ranges* | | | | | **Firing Side** | | **Opposite Side** | |
| Lined fire chamber | 30[1] | 762[1] | 36 | 914 | 24 | 610 | 18 | 457 |
| Unlined fire chamber | 30[1] | 762[1] | 36 | 914 | 36 | 914 | 18 | 457 |

[1]To combustible material or metal cabinets. If the underside of such combustible material or metal cabinet is protected with sheet metal of not less then 24 gauge [0.024 in (0.61mm)], spaced out 1 in. (25.4 mm), the distance shall be permitted to be reduced to not less than 24 in. (610 mm).
[2]Adequate clearance for cleaning and maintenance shall be provided.
[3]Provisions for fuel storage shall be located at least 36 in. (914 mm) from any side of the appliance.
[4]For clearances from air ducts, see NFPA 90B, *Standard for the Installation of Warm Air Heating and Air Conditioning Systems*.

Reprinted with permission from NFPA 211-2000, *Chimneys, Fireplaces, Vents, and Solid Fuel Burning Appliances*, Copyright© 2000, National Fire Protection Association, Quincy, MA 02269. This reprinted material is not the complete and official position of the NFPA on the referenced subject, which is represented only by the standard in its entirety.

## Table 4.2
## Methods for Reducing Clearances to Heat Producing Devices

| Clearance Reduction System Applied to and Covering All Combustible Surfaces within the Distance Specified as Required Clearance with No Protection (See 9-6.1.) | Maximum Allowable Reduction in Clearance (%) | | Where the required clearance with no protection is 36 in. (914 mm), the clearances below are the minimum allowable clearances. For other required clearances with no protection, calculate minimum allowable clearance from maximum allowable reduction.[9,10] | |
|---|---|---|---|---|
| | As Wall Protector (%) | As Ceiling Protector (%) | As Wall Protector (in.) (mm) | As Ceiling Protector (in.) (mm) |
| (a) 3½-in. (90-mm) thick masonry wall without ventilated air space | 33 | — | 24  610 | —  — |
| (b) ½-in. (13-mm) thick noncombustible insulation board over 1-in. (25.4 mm) glass fiber or mineral wool batts without ventilated air space | 50 | 33 | 18  457 | 24  610 |
| (c) 0.024-in. (0.61-mm), 24-gauge sheet metal over 1-in. (25.4-mm) glass fiber or mineral wool batts reinforced with wire, or equivalent, on rear face with ventilated air space | 66 | 50 | 12  305 | 18  457 |
| (d) 3½-in. (90-mm) thick masonry wall with ventilated air space | 66 | — | 12  305 | —  — |
| (e) 0.024-in. (0.61-mm), 24-gauge sheet metal with ventilated air space | 66 | 50 | 12  305 | 18  457 |
| (f) ½-in. (13-mm) thick noncombustible insulation board with ventilated air space | 66 | 50 | 12  305 | 18  457 |
| (g) 0.024-in. (0.61-mm), 24-gauge sheet metal with ventilated air space over 0.024-in. (0.61-mm), 24-gauge sheet metal with ventilated air space | 66 | 50 | 12  305 | 18  457 |
| (h) 1-in. (25.4-mm) glass fiber or mineral wool batts sandwiched between two sheets 0.024-in. (0.61-mm), 24-gauge sheets metal with ventilated air space | 66 | 50 | 12  305 | 18  457 |

1. Spacers and ties shall be of noncombustible material. No spacers or ties shall be used directly behind appliance or conductor.

2. With all clearance reduction systems using a ventilated air space, adequate air circulation shall be provided as described in 9-6.2.4. There shall be at least 1 in. (25.4 mm) between the clearance reduction system and combustible walls and ceilings for clearance reduction systems using a ventilated air space.

3. Mineral wool batts (blanket or board) shall have a minimum density of 8 lb/ft³ (128.7 kg/m³) and have a minimum melting point of 1500°F (816°C).

4. Insulation material used as part of clearance reduction system shall have a thermal conductivity of 1.0 (Btu-in.)/(ft²-hr-°F) or less. Insulation board shall be formed of noncombustible material.

5. If a single-wall connector passes through a masonry wall used as a wall shield, there shall be at least ½ in. (13 mm) of open, ventilated air space between the connector and the masonry.

6. There shall be at least 1 in. (25.4 mm) between the appliance and the protector. In no case shall the clearance between the appliance and the wall surface be reduced below that allowed in the table.

7. Clearances in front of the loading door or ash removal door, or both, of the appliance shall not be reduced from those in Section 9-5.

8. All clearances or thicknesses are minimums; larger clearances and thicknesses shall be permitted.

9. To calculate the minimum allowable clearance, the following formula can be used: $C_{pr} = C_{un} \times (1 - R/100)$. $C_{pr}$ is the minimum allowable clearance, $C_{un}$ is the required clearance with no protection, and R is the maximum allowable reduction in clearance.

10. Refer to Figures 9-6.2.1 (e) and 9-6.2.1(f) for other reduced clearances using materials found in (a) through (h) of this table.

NFPA 211, 1996 Edition

**Conclusion**

Several fires in recent history show how building systems and services affect fire growth and development as well as life safety. The most notable incidents include the World Trade Center Bombing in February of 1993, the One Meridian Plaza Building fire in Philadelphia in 1991, and the First Interstate Bank Building fire in Los Angeles in 1988. In each of these incidents, fire or smoke penetrated openings in floors or entered shafts and moved to areas remote from the initial fire. In the First Interstate fire, the fuel load provided by telecommunications equipment and workstations in the area of origin contributed to the fire's behavior. The use of unprotected elevators by staff early in the incidents resulted in a fatality at the First Interstate fire and an injury at One Meridian Plaza.

The systems and services in any building can provide the heat and/or fuel that results in a fire. This potential will have to be evaluated by the fire investigator. Beyond these direct contributions, the investigator must consider additional factors related to fire development and life safety. These factors include:

- The contribution of vertical shafts or horizontal openings to the spread of smoke or fire due to improper design or penetrations made after construction

- The contribution of the components of a building system as fuel packages that become involved after ignition

- Lack of proper fire-stopping of barrier penetrations by system components such as wire or pipe

- The use of heat-producing or flame-producing equipment during installation, maintenance, or repair of a system

- The potential for leaks in systems carrying ignitable liquids or gases in a building

Systems installed in today's buildings can be very complex and, in many cases, interconnected. The fire investigator must, as part of the investigation, determine whether a system played a part in ignition, growth, or spread of a fire. The investigator may have to call on outside experts who are familiar with the type of system in question due to its complexity.

# Chapter 5

# Basic Electricity for the Fire Investigator

**Performance Objectives**

This chapter provides information that addresses performance objectives described in NFPA 1033, *Standard for Professional Qualifications for Fire Investigator*, particularly those referenced in the following sections:

**Chapter 3 Fire Investigator**
**3-2.8**

(**NOTE:** This chapter references the *National Electrical Code®* (NEC) published by the National Fire Protection Association and widely adopted in the United States and more recently in Mexico and Central America. Part 1 of this standard provides safety standards for installation, and Part 2 establishes the product standards such as those developed by Underwriters Laboratories and referenced in the NEC. In Canada, fire investigators should refer to the *Canadian Electrical Code*, published by the Canadian Standards Association.)

Electricity is one of the basic commodities of modern society. Electrical energy is used to light buildings and operate most of the systems, appliances, and machines that we use in buildings. Fire investigators are likely to find electrical wiring, appliances, or machinery involved in many of the fires they investigate.

Under some circumstances, electricity and the wiring used to carry it can pose an ignition hazard wherever it is used. However, the investigator should understand that the mere presence of electrical equipment or wiring in or near the area of origin does not indicate that the ignition source was electrical. All factors must be evaluated equally before the source of ignition is established. The purpose of this chapter is to give the fire investigator an understanding of electrical terminology, basic electricity, electrical components, and typical wiring methods used to provide electricity in buildings and outside areas. It also provides the investigator with a knowledge base in electricity that will assist in the evaluation of potential electrical ignition sources.

This chapter begins with an overview of the basics of electricity. Voltage, current, power, and energy are addressed, as well as resistance and Ohm's Law. Simple circuits are used to illustrate these concepts.

##  Basics of Electricity

Electricity is the most widely used source of energy in today's society. (**NOTE:** Chapter 2, "Fire Behavior," discussed the terms energy, work, and power. These terms, with the same meanings, are frequently used in this chapter.) In its most basic form, *electricity* can be defined as the flow of electrons through a conductor. *Electrons* are small, negatively charged particles that revolve around the nucleus of atoms. Atoms make up all matter. Because electrons are loosely attached to the atom, under the right conditions they can be stripped off and moved to another atom. This results in a change in the "charge" of the atoms involved. This movement of charge in a conductor is an electric current. Materials with electrons that are free to move from one atom to another are called *conductors*. While copper and aluminum are the most common conductors used in wiring, most metals will also conduct electricity. Materials with atomic structures that do not allow the easy movement of electrons are not good conductors and are called *insulators*. Glass, plastics, rubber, and stone are examples of insulators.

While the actual particles (electrons) move very short distances, their movement causes a chain reaction of movements within the conductor. This movement is very rapid. When a charge enters a conductor at one end, it moves through it from atom to atom at the speed of light or almost instantaneously. For a charge to move in a conductor, it must have an energy source that causes the charge to move initially and a path for it to travel through. In discussing electricity, the path of travel for the charge is commonly called a *circuit*. The source of energy in a circuit can be a battery (chemical energy) or a generator (mechanical energy). The electrical force that causes the flow is called *voltage*.

An often-used analogy of water being circulated under pressure provides a helpful way to look at how an electrical circuit functions. The analogy illustrates an electrical system and serves as a reference for the definitions developed in the chapter (Figures 5.1 a and b).

## Basic Circuits

The simple circuit described in the previous analogy can be used to expand the discussion of electrical circuits. To relate the circuit to something most of us know about, we will build a circuit that could

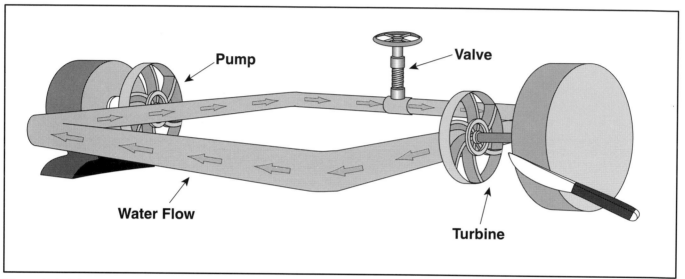

**Figure 5.1a** The flow of water is similar to the flow of electricity.

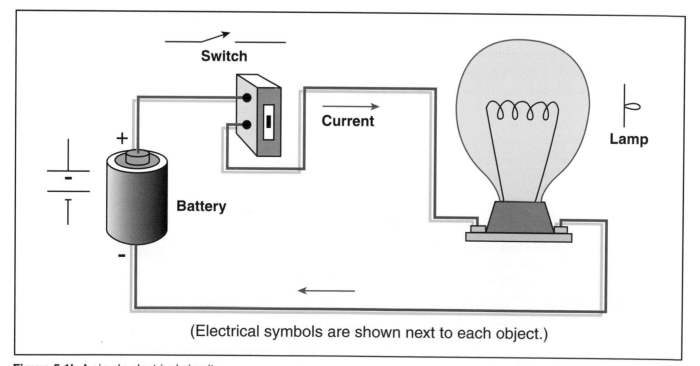

(Electrical symbols are shown next to each object.)

**Figure 5.1b** A simple electrical circuit.

Table 5.1 provides comparisons between the two systems shown in Figure 5.1. In this analogy, the pipe making up the water system and the wire conductors making up the electrical circuit are similar in that they both provide a pathway for a flow of water or electricity, known as *current* and expressed in *amperes* (*A*). A pump is used as a source of energy to force the liquid through the pipes in the hydraulic system. In an electrical system, a generator or battery provides the energy source. The pressure developed in the hydraulic system is measured in pounds per square inch (kPa). Electrical pressure is called *voltage* or *electromotive force* and is expressed in *volts* (*V*).

In both systems, the flow of water or electric current can provide the energy to do work. In the hydraulic system, the pipe can be arranged to pass through a turbine and turn a shaft that, in turn, can run machinery. In the electrical system, the current can be passed through a device or component, such as a bulb, that converts the electrical energy to light and heat energy or a motor that converts the electrical energy to mechanical and heat energy that can be used to run machinery. In both examples, the pressure or voltage will be reduced after performing the work. This drop is measured with pressure gauges in the hydraulic system and by a voltmeter in the electrical system. In a hydraulic system, the flow may be regulated using a valve, and in an electrical circuit, current is controlled with a switch. Note also that in these examples, the flow of liquid or electricity is in one direction. This is analogous to a direct current (DC) circuit. In an alternating current (AC) circuit, the direction of flow would be constantly reversing.

As the liquid flows through the hydraulic system, there is a pressure loss due to the friction of the pipe. This loss of pressure is commonly known as *friction loss*. As electrons flow in a conductor, there is a similar loss of voltage. The friction in a conductor or other component in an electrical circuit is called *resistance*. The resistance or friction loss produces heat in both systems. Finally, in both systems, the amount of friction loss or resistance in the system is directly related to the size of the conductor (pipe or wire), the length of the conductor, and the material from which it is made. The larger the cross-sectional area of the conductor, the less resistance there is to the flow within it. The longer the conductor is, the more resistance it has to the flow in it. In pipe, the material and how smooth it is on the inside affects the resistance to flow. In wire, the material from which it is made affects the resistance. For instance, a length of copper wire has less resistance than the same length of aluminum wire because of its physical properties.

## Table 5.1
## Water and Electrical System Comparison

| Elements of the Water System | Elements of the Electrical System |
| --- | --- |
| Pump | Generator |
| Pressure | Voltage |
| Pounds per square inch (psi/kPa) | Volts (V) |
| Pressure gauge | Voltmeter |
| Water | Electrons |
| Flow | Current |
| Gallons per minute (gpm) | Amperes (A) |
| Flowmeter | Ammeter |
| Valve | Switch |
| Friction | Resistance (ohms) |
| Friction loss | Voltage drop |
| Pipe size inside diameter | Wire size—AWG No. |

Reprinted with permission from NFPA 921-1998, *Guide for Fire and Explosion Investigations,* Copyright© 1998, National Fire Protection Association, Quincy, MA 02260. This reprinted material is not the complete and official position of the NFPA on the referenced subject, which is represented only by the standard in its entirety.

be used to power the headlights on a vehicle. Figure 5.2 is essentially the same circuit used in the analogy shown in Figure 5.1, with a 12-volt battery for an energy source, two headlights, and a switch controlling the current from the battery to the lights. For the lights to operate, current must pass through the switch and both lights and back to the negative terminal of the battery. When the switch is opened (off), the lights are not illuminated because there is no current in the circuit. Now, examine the diagram and consider what would happen if one of the lights burned out with the switch in the *on* position.

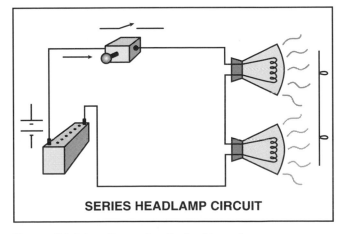

**SERIES HEADLAMP CIRCUIT**

**Figure 5.2** A headlamp circuit wired in series.

As wired, all of the current passes through the switch and both lights. Should one light burn out, it would have the same effect as opening the switch — removing the current from the circuit. Thus, if one light burns out, the other will not light. This configuration, where the current in a circuit flows through all components, is called a *series* circuit.

The use of a series circuit for the headlight circuit on the vehicle obviously would not be the best choice. For this important circuit, a configuration that would allow one light to light — should the other one burn out — would be better. Figure 5.3 shows such a circuit for the vehicle. In this configuration, each lamp is provided with a unique path for the current from the switch to the light. Should a light burn out now, the other light will still light as the current has a complete path to travel through. This configuration is called a *parallel* circuit. Parallel circuits are the most common circuits found today. The reason for using this configuration is that all components connected to a parallel circuit operate at the same voltage, and components may be added or removed from the circuit without interfering with the operation of other components (Figure 5.4).

Figure 5.5 shows several modifications to the circuit. In this circuit, the voltage comes from the positive side of the car's 12-volt battery to a fuse block, where it is distributed. The negative side of the battery is connected to the vehicle frame, creating a ground. In this arrangement, the vehicle frame becomes one of the conductors in the headlight circuit. (**NOTE:** Circuits in homes and other buildings do not use conducting materials, such as structural steel or metal water pipes, as a return path for current. A separate conductor — called the grounded, common, or neutral conductor — is always used.) From the fuse block, wire is used to connect the positive side of the battery to the control switch for the

headlights. In order to turn the lights on and off, a switch controls the circuit. From the switch, a separate power wire is connected to each light. The opposite side of the light is connected to the vehicle frame or ground to complete the circuit back to the battery. When the switch is open, the circuit is not complete, and the lights are off. When the switch is closed, the circuit is completed and there is current from the battery to the lights and back to the battery. The current through the lights causes them to illuminate, thus giving off light and heat energy.

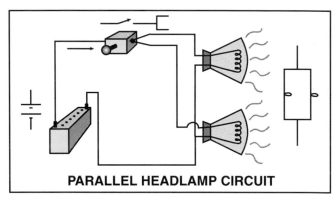

**PARALLEL HEADLAMP CIRCUIT**

**Figure 5.3** A headlamp circuit wired parallel.

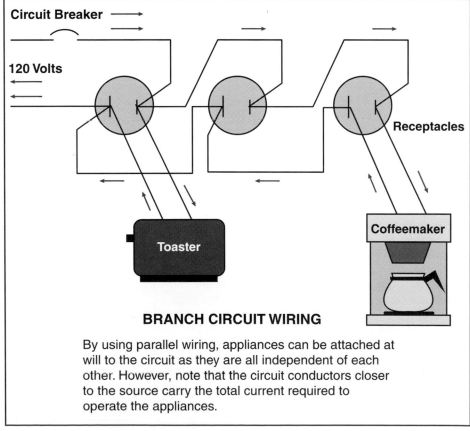

**BRANCH CIRCUIT WIRING**

By using parallel wiring, appliances can be attached at will to the circuit as they are all independent of each other. However, note that the circuit conductors closer to the source carry the total current required to operate the appliances.

**Figure 5.4** A simple branch circuit diagram.

Compare the circuit in Figure 5.5 to the simple circuit shown in Figure 5.1. The 12-volt battery is the energy source for the circuit. That is, the battery is the source of the energy that results in the electrons being pushed through the circuit resulting in *current*. To measure voltage in the circuit, a voltmeter is placed across the circuit as shown in Figure 5.6. To measure current in the circuit, an ammeter is placed in the line so that the current flows through it.

As current passes through the circuit, it encounters some opposition from the wire and components in the circuit. This opposition to flow is called *resistance* and is measured in *ohms* (W). In our example, the wire that makes up the circuit and the lights in the circuit all have resistance. The size and composition of the wire used in a circuit will have some effect on the total resistance found in the circuit. Table 5.2 provides the resistance per 1,000 feet (300 m) of wire for three common conductor sizes of both copper and aluminum.

Wire size is given in American Wire Gauge (AWG) increments and is determined by the diameter of the wire in question. The numbers will increase as the diameter of the wire decreases (e.g., No. 14 AWG wire is smaller than No. 8 AWG wire). The lower AWG rating has less resistance and is therefore able to carry a greater amount of current. As shown in Table 5.2, different materials have different resistance. Both copper and aluminum are considered to be good conductors in that they offer relatively low resistance to the flow of current in a circuit. However, of the two, copper is the better conductor because it has a lower resistance. Other materials, such as nichrome, are very poor conductors and offer very high resistance to current flow. Nichrome, in fact, is used as the heating element in many heat-producing appliances, such as space heaters and kitchen toasters.

If there is a total of 20 feet (6.1 m) of No. 12 AWG copper wire used in our automobile headlight circuit, the total resistance can be calculated in the

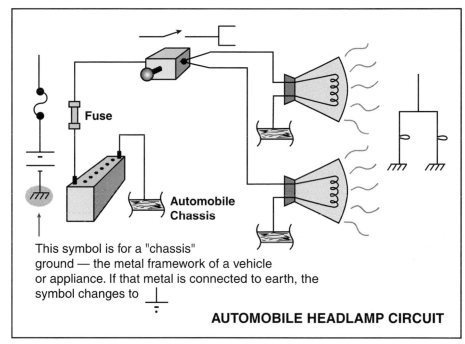

**Figure 5.5** An automobile headlamp circuit.

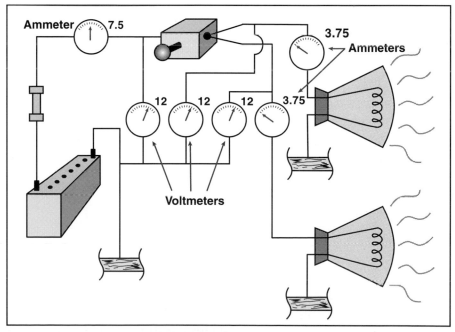

**Figure 5.6** Schematic of a typical voltmeter.

circuit from the wire. From Table 5.2 we find that 1,000 feet (300 m) of No. 12 AWG copper has a total resistance of 2 ohms, thus 1 foot (0.3 m) of the wire has a resistance of .002 ohms ($2\,\Omega/1{,}000\,\text{ft} = .002\,\Omega$). The resistance of the 20 feet (6.1 m) of wire in the circuit would then be $20 \times .002$ or $.04\,\Omega$. The resistance in the circuit from the wire is negligible (the resistance of a 45 watts headlight bulb is 3.2 ohms). This is the case in almost all circuits with the exception of those with very long runs of wire, those that use very small gauge wire, or circuits with poor connections.

## Ohm's Law

In any circuit, the relationship between resistance, current, and voltage can be stated mathematically using Ohm's Law, where:

Voltage = Current x Resistance

The units used in computations of Ohm's Law are shown in Table 5.3.

Ohm's Law can be used to calculate the voltage, current, or resistance in a circuit if the other two values are known. Figure 5.7 provides the equations for all three values and a common diagram used to help remember the equations.

Using the headlight circuit, Ohm's Law illustrates the relationship between voltage, current, and resistance. We know that the voltage in the circuit is 12 volts. If the circuit branch to one light has a total resistance of 3.2 ohms, the current in that part of the circuit would be calculated as follows:

$I = 12\,\text{V}/3.2\,\Omega$

$I = 3.75\,\text{A}$

Another example might be the toaster in your kitchen. While the appliance uses AC power, the basic calculations are the same. From the manufacturer's data, it is determined that the resistance of the heating element during operation is 10 ohms and the operating voltage is 120 volts. Ohm's Law allows us to calculate the current the appliance will draw as follows:

$I = 120\,\text{V}/10\,\Omega$

$I = 12\,\text{A}$

## Power

When current moves through a circuit, energy is spent. This energy/power expenditure may appear in any number of ways, including light coming

**Table 5.2**

| Size of Wire AWG | Resistance in Ohms Copper | Resistance in Ohms Aluminum |
|---|---|---|
| 14 | 3.1 | 5.1 |
| 12 | 2.0 | 3.2 |
| 10 | 1.2 | 2.0 |

Source NFPA 70, *National Electrical Code ®*, Table 8, Chapter 9
AWG = American Wire Gauge

Resistance values per 1,000 feet (300 m) of wire at 167°F (75°C)

**Table 5.3**

| | Symbol | Unit |
|---|---|---|
| Voltage | V | volt (V) |
| Current | I | ampere (A) |
| Resistance | R | ohm (Ω) |
| Power | P | watt (W) |

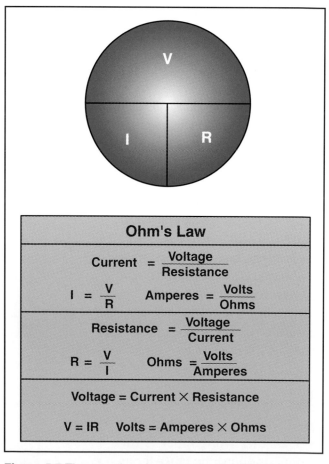

**Figure 5.7** The equations involved in using Ohm's Law.

from the headlights, the vehicle horn sounding, the starter motor in the vehicle spinning, or toast being made in a toaster. In our discussion in Chapter 2, *power* was defined as the amount of energy delivered over a given period of time. In electrical circuits, the unit used for power is the *watt* (*W*). These units should be familiar because most electrical devices provide a wattage rating. For DC circuits and typical household AC circuits, power is calculated using the following formula:

$$P = VI$$

By using Ohm's Law, the following relationships can be developed and applied, depending on which of the quantities is known by the investigator:

$$I = P/V$$
$$V = P/I$$
$$P = I^2R$$
$$P = V^2/R$$

Using the headlight circuit example, it is known that the voltage is 12 volts and that the current in the portion of the circuit leading to a single light is 3.75 amperes when the light is on. (**NOTE:** The .04 ohms resistance of the wire is so small in comparison to the resistance of the headlight that it is ignored in the calculation.) With this information, the power consumption of one of the lights can be calculated as follows:

$$P = VI$$
$$P = 12 \text{ V} \times 3.75 \text{ A}$$
$$P = 45 \text{ W}$$

The total power consumption in the headlight circuit would be the sum of the wattages of the two lights when operating, or 90 watts. The combined wattages of the components in any circuit make up the *load* on that circuit. Looking at Figure 5.3, you could calculate the current for this load in the portion of the circuit prior to the split at the switch using the equation for current as follows:

$$I = P/V$$
$$I = 90 \text{ W}/12 \text{ V}$$
$$I = 7.5 \text{ A}$$

Using the kitchen toaster discussed earlier, we can calculate its power consumption:

$$P = VI$$
$$P = 120 \text{ V} \times 12 \text{ A}$$
$$P = 1,440 \text{ W}$$

To determine the total load on a circuit, the fire investigator just has to add up the wattage of the individual components or appliances connected to a circuit. With this information, Ohm's Law can be used to calculate any of the other variables in the circuit, such as total current or resistance.

The common branch circuits found in most homes have either a 15- or 20- ampere capacity. The cable used will be No. 12 AWG copper or No. 10 AWG aluminum for 20-ampere circuits and No. 14 AWG copper or No. 12 AWG aluminum for 15-ampere circuits. Most modern installations will use copper in branch circuits. Using the equation for power, the load capacity for these circuits can be calculated.

$$P = VI$$
$$15 \text{ A} \times 120 \text{ V} = 1,800 \text{ W}$$
$$20 \text{ A} \times 120 \text{ V} = 2,400 \text{ W}$$

The fire investigator can determine the load on a branch circuit by simply adding up the wattages of the components connected to the circuit (Table 5.4). When doing this, however, the investigator should keep in mind that the wire sizes for electrical wiring used in branch circuits in buildings and other locations are determined by the requirements

| Table 5.4. Power Consumed by Typical Appliances | |
| --- | --- |
| **Appliance** | **Watts** |
| Air conditioner (room type) | 800-1500 |
| Electric blanket | 175 |
| Coffeemaker | 600 |
| Dishwasher | 1800 |
| Hair dryer | 1500 |
| Clothes dryer | 4500 |
| Portable heater | 1200 |
| Household iron | 1000 |
| Electric motor (per hp) | 1000 |
| Built-in oven | 4000 |
| Electric stove (all burners and oven on) | 8000-16000 |
| Refrigerator | 250 |
| Television | 250 |
| 80 gallon electric hot water heater | 4500 |

of NFPA 70, *National Electrical Code®*. These requirements are based on the size of the wire, the material from which it is made, and the temperature at which it is intended to operate. Minimum wire sizes in the code are very conservative and have a significant safety factor built into them. Simply finding a circuit that appears to be overloaded may not cause sufficient heat production to result in an ignition.

## Alternating Current

In the hydraulic system analogy presented early in the chapter, the concept of direct and alternating current was introduced. In *direct current* (DC), electrons flow in only one direction — from negative to positive. Our example, using an automobile headlight circuit, is a direct current circuit with energy supplied by a battery.

The second type of current is called alternating current (AC). AC is the common type of current found in the electrical systems in most buildings. Thus, it is the type that the fire investigator will see most frequently when investigating building fires. One of the few exceptions that will be found in

today's buildings is the source of energy for elevators — they are often operated using DC power.

In *alternating current*, the direction of the flow of electrons is constantly changing. Alternating current is a function of the method used to produce the energy. Instead of a battery used for the headlight example, AC is produced using a generator. A *generator* is basically a coil of conductors that is rotated within a magnetic field. As the coil is turned in the field, a voltage is produced. As the coil turns, the voltage and its charge "polarity" is constantly changing. Figure 5.8 shows a coil turning in a field and a graph of the resulting voltages. In mathematics, this type of curve is called a *sine curve* or *wave*. As you can see in the figure, the voltage goes from 0 to maximum as the coil turns to 90 degrees, then the polarity changes, and the voltage returns to 0 at 180 degrees. The maximum voltage of the opposite polarity occurs at 270 degrees and then returns back to 0 at 360 degrees. One complete turn of 360 degrees is called a *cycle*. The number of cycles that occur in one second is called the *frequency*, which is expressed in *hertz* (Hz). The AC power used in North America has a frequency of 60 cycles per

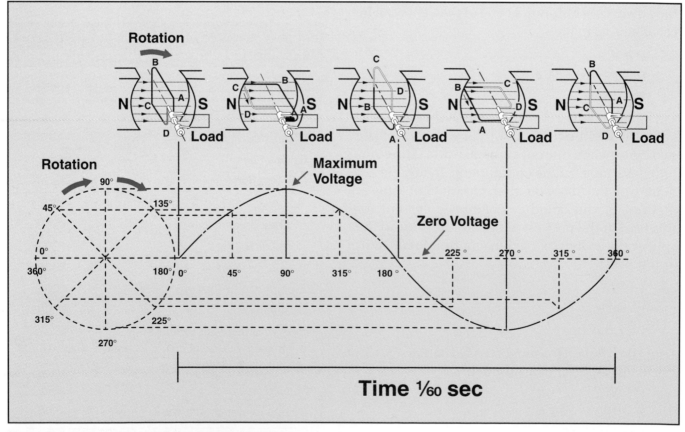

**Figure 5.8** Voltage rises and falls as the coil rotates.

second or 60 Hz. The most common voltages are approximately 120 volts and 240 volts. Other voltages may be seen in some industrial or special properties.

Electric companies supply alternating current to their customers because it is the most efficient and versatile method of producing and distributing electrical energy. Electric generators that are turned by turbines powered by water or steam can produce energy very efficiently. When the electrical energy is produced, it can be transmitted at high voltages. Higher voltages help in the distribution of electricity because the higher the transmission voltage, the lower the current in the lines. The resistance of the long transmission lines results in a loss of energy, primarily due to the electrical energy being converted to heat. The power equation **P = I²R** shows that as the current in a transmission line is decreased, the amount of power lost due to resistance also decreases. This is accomplished by using high voltages. The power loss due to resistance increases significantly as the transmission voltage is decreased.

Although power is distributed at high voltages, the voltage must be reduced before residential and commercial customers can use it. This is accomplished using step-down transformers. Transformers operate using primary and secondary coils that are wrapped around an iron core. As the alternating current passes through the primary coil, a magnetic field is created. The rapid variation from positive to negative and back creates a voltage in the secondary coil. Depending on the ratio of windings in the primary and secondary coils, the voltage created in the secondary coil will be higher (step-up) or lower (step-down) than the

voltage in the primary coil. The most common transformer found in the power distribution grid is the step-down transformer. Figure 5.9 provides an example of a very simple step-down unit.

### Example of Power Loss in Transmission

The following is an example of the energy loss that would occur in transmitting 10,000 watts of electrical power over a No. 1 AWG cable for one mile (1.6 km) at voltages of 120 volts and 10,000 volts.

First, calculate the current using Ohm's Law.

The current required for 10,000 W at 10,000 V would be:

$$I = P/V$$
$$I = 10,000 \text{ W}/10,000 \text{ V}$$
$$I = 1 \text{ A}$$

The current required for 10,000 W at 120 V would be:

$$I = P/V$$
$$I = 10,000 \text{ W}/120 \text{ V}$$
$$I = 83 \text{ A}$$

From the NEC, we find that one mile (1.6 km) of No. 1 AWG wire has a resistance of .665 Ω. Using the power equation, we can calculate the power loss due to resistance at the two voltages.

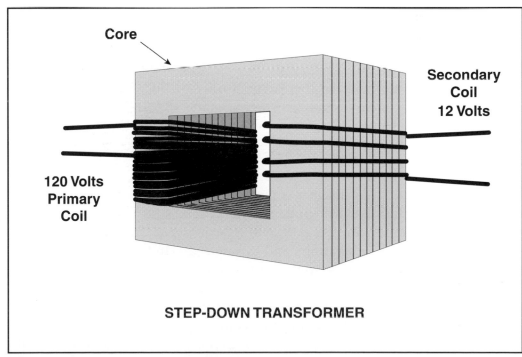

Figure 5.9 A typical step-down transformer.

For the transmission of the power at 10,000 V, the current was 1 A and the line loss is:

$P = I^2R$

$P = (1A)^2 \times .665\ \Omega$

$P = .665\ W$

For the transmission of the power at 120 V, the current was 83 A and the line loss is:

$P = I^2R$

$P = (83\ A)^2 \times .665\ \Omega$

$P = 4,580\ W$

The transformer shown in Figure 5.9 has a primary winding of two loops with a one loop secondary. If 10 volts were applied to the primary, the resulting voltage in the secondary would be 5 volts. One additional concept the fire investigator should keep in mind when learning about transformers is that the power into and out of a transformer is the same. This is based on the concept of the Conservation of Energy previously discussed in Chapter 2, "Fire Behavior." The only reduction in power in a transformer is that which is converted to another form of energy, such as heat due to resistance. If the simple transformer in our example was 100 percent efficient and 1 ampere of current entered the primary coil, the current in the secondary would be 2 amperes. This is illustrated again with the power equation. In the primary winding the power would be:

$P = VI$

$10\ V \times 1\ A = 10\ W$

Because the energy is conserved, there must also be 10 watts in the secondary winding. The current would be calculated using the power equation:

$I = P/V$

$10\ W/5\ V = 2\ A$

This discussion of transformers is provided to give the fire investigator a basic understanding of the function of the transformer. These units will be found in electrical substations and throughout the power distribution system in our communities (Figures 5.10 a and b). Transformers may also be located as part of the facilities for occupancies that require a large amount of power to operate, such as health care institutions and commercial and industrial complexes. They reduce commercial/industrial voltages of up to 600

**Figures 5.10 a and b** Two of many types of transformers.

volts to typical 120/240 volt levels. The investigation of fires in such equipment will more than likely require that the fire investigator seek expert assistance to complete the investigation related to the electrical system. Fire investigators will also encounter step-down transformers used to reduce normal 120 volt electricity to lower voltages (12 V, 9 V, 4.5 V, or others) used to power appliances or devices.

 ## Common Electrical Systems in Buildings

In most jurisdictions in the United States, the installation of electrical systems is regulated by NFPA 70, *National Electrical Code®* (NEC). In Canada, the document used is the *Canadian Electrical Code*. The fire investigator can find additional information regarding electrical systems and their components in these documents. The electrical code is an essential reference for the investigator examining a fire scene where there is a possibility that the electrical system was involved in the ignition of the fire. The following discussion provides the investigator with an overview of a code-compliant electrical system in a residential property.

## Service

Electricity enters the building via a service (Figure 5.11). The service may be aboveground (service drop), but more and more services are being run underground (service lateral). The service wires will be at least three conductors that lead from the power company's lines to the building. The NEC requires that these conductors be a minimum of No. 8 AWG copper or No. 6 AWG aluminum. The size of the service conductor will be determined by the ampacity of the service. Most modern service connections to residential properties are made with three No. 2 AWG aluminum conductors. Two of the cables will be insulated with the third uninsulated conductor serving as a grounded neutral conductor. There will be a voltage of 120 volts between either of the insulated conductors and the bare one and 240 volts between the insulated conductors.

The service lines are run into the building in such a way as to prevent damage or unintentional contact with the conductors. Aboveground cables will be securely attached to the structure at the service

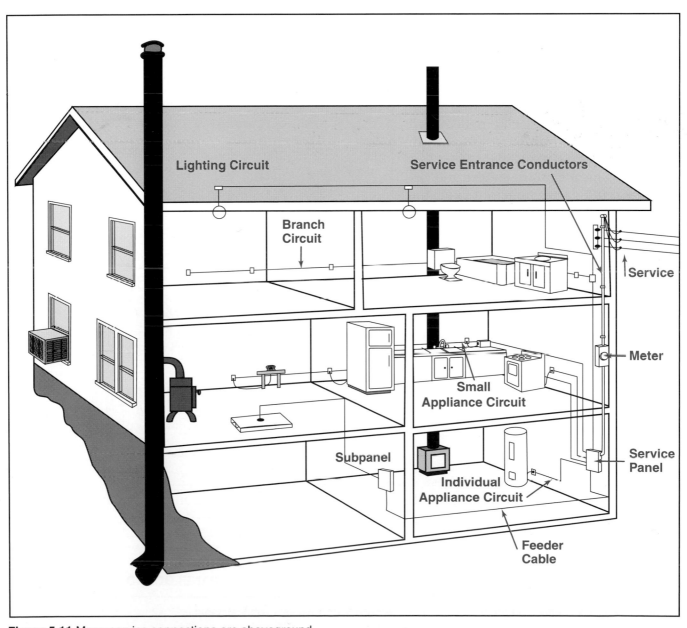

**Figure 5.11** Many service connections are aboveground.

entrance. From the service entrance, the line will run to a meter box that houses a meter used to determine the consumption of energy by the customer. The meter measures the total energy flow into the building in kilowatt-hours (kW-h) and is in place for billing purposes. From the meter, the conductors will be brought into a panelboard where the power will be distributed to the electrical system (branch circuits) in the building. At some point between the meter and the branch circuit protective devices, there will be a disconnect for the service. This disconnecting means may be located in the panelboard (circuit breaker/fuse panel) or a separate box. Article 230 of the NEC allows the service disconnect(s) to be up to six separate switches or circuit breakers (Figure 5.12).

To use the power in a building — in this case a home — the service is brought into a panelboard where it is distributed to branch circuits within the building. In most installations, the main disconnecting means and the overcurrent protection for the branch circuits are located in the same panelboard. This panel is commonly called the *circuit breaker panel*. Figure 5.13 is an example of a typical panelboard with a disconnecting means for the service and circuit breakers for each branch circuit in the building.

Power for mobile homes is provided using a feeder assembly that is made up of a four-wire cable that attaches to the mobile home with an attachment plug cap. The capacity of the plug cap is normally 50 amperes. In these installations, grounding and grounded (neutral) conductors are separated past the metering point.

## Branch Circuits

*Branch circuits* are those circuits that lead from an overcurrent protection device in the circuit breaker panel to service the rooms and equipment in a building. The branch circuits in a building are used for lighting, the operation of nonfixed electrical equipment (i.e., appliances that are attached to the system via a plug attached to an electrical receptacle), and devices that are permanently attached to the circuit that feeds them, such as fixed cooking equipment in a home. Figure 5.14 shows the various branch circuits in a typical home.

In the typical residential electrical system installed to today's standards, several 15- or

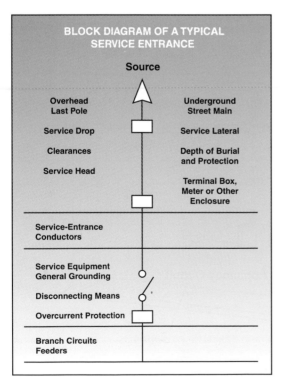

**Figure 5.12** Typical service entrance components.

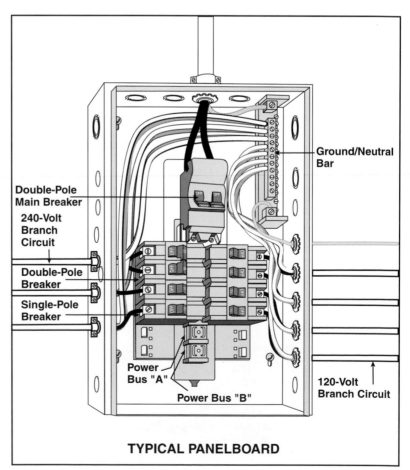

**Figure 5.13** A typical electrical panel.

# TYPICAL RESIDENTIAL ELECTRICAL SYSTEM

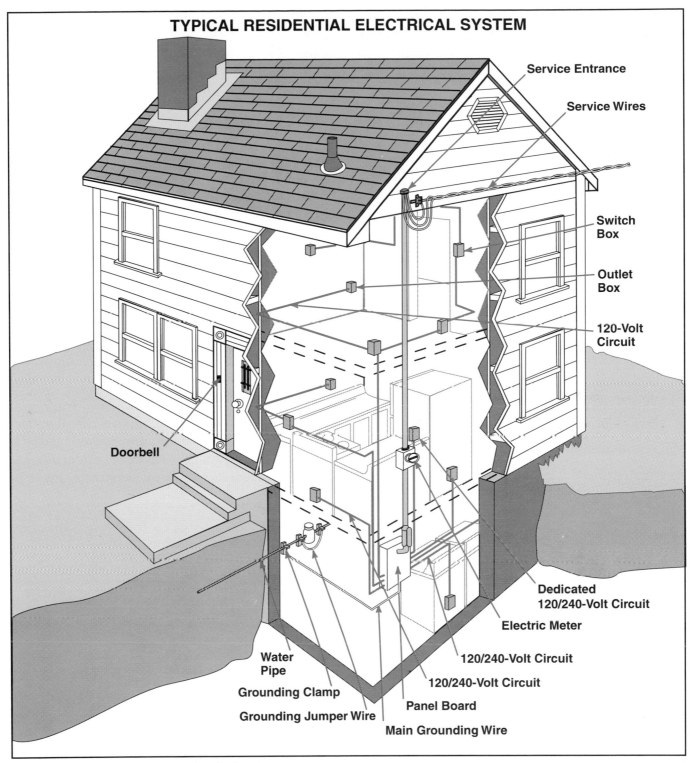

**Figure 5.14** A typical residential electrical system.

20-ampere general purpose branch circuits will be provided in most rooms for lighting and to supply wall receptacles. Additional 20-ampere branch circuits will be provided as appliance circuits in the kitchen, bathrooms, and in the laundry area. There may also be several individual 30-ampere branch circuits that supply power directly to a specific device or appliance, such as an electric water heater, a clothes dryer, garbage disposal, or water pump for a well. Circuits rated at 40 or 50 amperes are permitted in dwellings to supply fixed cooking appliances. These circuits will have only one heavy-duty receptacle located near the rear of the stove unit. For workshops or other locations where there may be a heavy power demand, subpanels fed by feeder lines from the panelboard are permitted. Table 5.5 gives an overview of the type and size of these circuits in residential installations.

## Circuit Wiring

The NEC defines the acceptable materials and methods for the wiring of branch circuits in a building. The most common cable found in residential properties is nonmetallic shielded cable. One popular brand is Romex™. These cables are factory assemblies with two or more wires, and they have a nonmetallic outer sheath that is moisture resistant and fire retardant. This type of cable is capable of being installed directly or fished into partitions and other spaces. The NEC does not permit nonmetallic sheathed cable in the following locations:

- Dwelling or structure that exceeds three stories above grade

- Service entrance cable

- Commercial garages having hazardous locations

- Theaters or similar locations

- Motion picture studios

- Storage battery rooms

- Hoistways

- Embedded in poured cement, concrete, or aggregate

- Hazardous locations (as defined in Article 500 of the NEC)

All conductors serving branch circuits must be protected from physical damage. The NEC allows cable to be run on a surface, through bored holes and notches in wood framing, or through holes in metal framing that are protected with bushings or grommets. When the holes for cable are located where there is a potential for nails to hit the cable, the NEC requires that metal plates be provided for protection. Other than in one- and two-family dwellings, branch circuits will be protected using conduit, cable raceways, or by some other approved means. Specific wiring methods and requirements are found in Article 300 of the NEC.

Section 300-15 of the NEC requires that all splices and connections be made in approved electrical boxes. Thus, any splice in a wire or connection to a device, such as a switch or receptacle, must be

| | Table 5.5 Branch Circuit Requirements | | | | |
|---|---|---|---|---|---|
| Circuit Rating | 15 Amp | 20 Amp | 30 Amp | 40 Amp | 50 Amp |
| Min. conductor size | 14 AWG | 12 AWG | 10 AWG | 8 AWG | 6 AWG |
| Outlet devices | Any type | Any type | Heavy duty | Heavy duty | Heavy duty |
| Receptacle rating | 15 amp | 15 or 20 amp | 30 amp | 40 or 50 amp | 50 amp |
| Type of load permitted | Lighting or utilization equip. | Lighting or utilization equip. | Utilization equip. | Fixed cooking equipment | Fixed cooking equipment |
| Overcurrent protection | 15 amperes | 20 amperes | 30 amperes | 40 amperes | 50 amperes |

Note: 30-ampere and greater branch circuits may be used for lighting circuits provided with heavy duty lamp holders in occupancies other than dwelling units.

Note: Utilization equipment is defined in NFPA 70 as "equipment that utilizes electric energy for electronic, electromechanical, chemical, heating, lighting, or similar purposes."

Source: Table 210-24, NFPA 70, *National Electric Code*

made in an electrical box. The size of the box required is determined by the size and number of conductors that enter the box. The conductors entering a box must be secured with a clamp or other approved means at the point of entry.

Within the electrical box, the actual connections are made by using terminal screws on a device, such as a switch or receptacle, or with an approved connector such as a wire nut. Connections are a very important part of the electrical system, as they are a common failure point. Poor connections can result in increased resistance and thus increased heat that can, under the right conditions, cause the ignition of nearby combustibles. Proper connections are tight and made using approved devices and methods. A point of special interest is a connection made with aluminum wire. For this application, devices specifically approved for aluminum must be used. These devices are marked with AL/CU, and older devices are marked with CO/ALR. The joining of copper and aluminum wire directly together is prohibited. Devices approved for splicing copper to aluminum provide for the separation of the two metals in the device.

## Grounding

The grounding of an electrical system is required to reduce the hazard of electrical shock. The system is grounded to protect against lightning, accidental contact of high-voltage conductors with lower voltage conductors, and any contact between an energized conductor and grounded objects. The ground wire provides a low resistance path for the current to follow in the event of an accidental contact. This flow will normally be sufficient to cause the circuit protection to operate and stop the current flow in the branch containing the fault.

In modern construction, the entire electrical system — from the service entrance through all of the branch circuits — is grounded. Fire investigators may find 2-wire, ungrounded, branch circuits in older homes. In newer systems, one of the three connectors that provide the service will be a ground wire (normally bare wire). The panelboard is then grounded by connecting to the electric utility neutral (grounded) conductor — and as a backup — to a grounding rod driven into the ground near the service entrance and/or metal water piping systems. From the circuit breaker in the panelboard, a modern branch circuit is grounded throughout. The grounding (bare or green) conductors leave the panel from the ground bus and go to the boxes that make up the circuit. At each box, the grounding conductor is attached to the box, if it is metal, and then to the grounding wires of the cables leading into and out of the box (Figure 5.15). Recall that although the neutral (grounded white or neutral gray) conductor is not energized, it does carry current and remains isolated from the grounding conductor except at the panelboard.

The grounding of devices that operate using electricity is provided for the protection of those using the device. If the insulation on the internal wiring were to break down or be damaged and come into contact with a metal case or housing, a severe shock could occur. The case or housing of the appliance is connected to a grounding conductor in the cord. The ground conductor is then connected to a three-prong receptacle on a branch circuit. The ground conductor from the appliance is in contact with the system ground at this point. Should the case become energized, a short circuit develops between the case and the ground, causing the overcurrent protection on the circuit to operate.

It should be noted that while defects in grounding may lead to shock dangers, they do NOT result in an increase in fire danger. Of course, poor grounding may indicate a poor quality wiring installation in general.

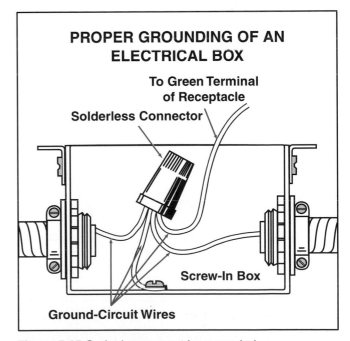

**PROPER GROUNDING OF AN ELECTRICAL BOX**

To Green Terminal of Receptacle

Solderless Connector

Screw-In Box

Ground-Circuit Wires

**Figure 5.15** Outlet boxes must be grounded.

## Circuit Protection

The panelboard provides a point at which the power entering a building is divided into the branch circuits. At the origination point of the branch circuit, some type of overcurrent protection is provided. In modern residential installations, this is normally a circuit breaker. However, the fire investigator may find the circuits protected with fuses in commercial/industrial and in older residential systems. The NEC requires overcurrent protection for both equipment and wiring in an electrical system. Overcurrent protection is provided to protect against excess current in a circuit. Excess current in a circuit can result from a short circuit or the overloading of the circuit with a load that is in excess of the rated capacity.

A *short circuit* is a low-resistance path between conductors that allows a high current flow. Applying Ohm's Law to a kitchen toaster in an earlier example, it was found that with a 10 ohms resistance, the device would draw 12 amperes. Reapply Ohm's Law for a much smaller resistance that could occur as a result of someone damaging the insulation of one of the conductors, allowing the energized conductor to come into contact with the neutral. In this example, the total resistance of all the components in the circuit might be 0.5 ohms. The current in the circuit would be:

$$I = V/R$$
$$I = 120 \text{ V}/0.5 \text{ W}$$
$$I = 240 \text{ A}$$

The reduction in resistance resulting from the short circuit results in a very significant increase in current flow in the circuit — much more than the conductors can safely carry. Short circuits can result from:

- Malfunctions in appliances or electrical components

- Improper wiring of branch circuits or components

- Physical damage to conductors that result in conductors firmly contacting each other (a bolted fault)

Overloads occur when components requiring more power than a circuit can safely carry are connected to a circuit. While circuits meeting the requirements of the NEC have a significant safety margin built into them, overloading a circuit stresses all components in the circuit. Before tripping a breaker or blowing a fuse, an overload condition could cause poor connections in a circuit to overheat and potentially ignite nearby combustibles. When the overload condition reaches the rating of the device, it will operate (circuit breakers trip, fuses blow) and open the circuit, removing the current.

The protective devices in the electrical circuit are there to prevent the overheating of conductors in the circuit. Should the protective device be sized incorrectly or not operate properly, there is potential for heat damage to the insulation on the conductor. Overcurrent protection that is properly sized and operating provides a level of safety and dependability that helps to prevent the electrical system from becoming an ignition source.

The fire investigator will see two types of devices for overcurrent protection: fuses and circuit breakers. A *fuse* is a device that provides protection to the circuit using a fusible element that melts from the heat generated by excessive current flow. The fire investigator will encounter three basic types of fuses in the field. The most familiar type is the common plug-type fuse that screws into a socket in the panelboard. There are two common plug-type fuses: Edison-based type and S type. These devices are still commonly found in older domestic and farm installations. Plug fuses are available in various ratings from 30 amperes at 125 volts down. Cartridge fuses are used in applications where protection in excess of 30 amperes is required (Figure 5.16).

Circuit breakers provide overcurrent protection and, unlike fuses, can be reset after they have tripped. Circuit breakers can also be used as a line switch for the circuit they serve. Most residential

**Figure 5.16** Typical cartridge fuses.

circuit breakers operate by both a thermal and magnetic means, depending on the type of fault to which they are exposed. A bimetallic strip is used to provide overload protection. The bimetallic strip is made of two dissimilar metals that are bonded together. The bimetallic strip is part of the circuit, and the total current in the branch circuit flows through it. The strip is calibrated to open the circuit when the amount of heat generated by the current flow exceeds the rating of the breaker. As it is exposed to heat, the strip bends — due to the different expansion rates of the two metals — and breaks the circuit. The operation of a circuit breaker in response to a short circuit is achieved using a magnet to activate the trip bar in response to the magnetic field created by the large overcurrent condition.

The time required for a breaker to trip depends on the amount of overcurrent. A typical 20-ampere breaker might take 15 minutes to trip with a current of 30 amperes, 30 seconds for 40 amperes, and less than 0.1 second for 200 amperes. This trip time versus amount of overcurrent is called a *trip curve* and is available from manufacturers of the devices.

Circuit breakers are designed so that they will trip even if the breaker handle is wedged into the *on* position. It should be noted that the thermal trip mechanism will operate in response to high ambient temperatures in the vicinity of the breaker; in some cases, even if the breaker is in the *off* position. Thus, if the fire is in close proximity to a panel, the fire investigator might find all breakers in the tripped position after the fire.

## Ground-Fault Protection

A *ground fault* is an unintended path of current flow between a conductor and ground. NFPA 921, *Guide for Fire and Explosion Investigations* (1998), defines a ground fault as:

A current that flows outside the normal circuit path, such as:

(a) Through the equipment grounding connector,

(b) Through conductive material other than the electrical system ground (metal water or plumbing pipes, etc.),

(c) Through a person,

(d) Through a combination of these ground return paths.

Because the resistance of the human body is so high, the current with a body in a circuit is usually well below the rating of the fuse or circuit breaker in the circuit. As a result, the normal overcurrent protection provided will not operate in these life-threatening conditions. The device used to protect against ground faults is the *ground fault circuit interrupter* (GFCI). This device operates very rapidly in response to small amounts of current leakage. The primary purpose of the device is to provide protection against electric shocks. GFCIs use electronic monitoring circuits to monitor the current flow in both the hot and neutral conductors supplying the load. If the circuit detects less current in the neutral conductor — an indication that current is being diverted (possibly through a person) — it trips and opens the circuit, stopping the current. While fuses and circuit breakers are provided for the protection of the conductors in a circuit, GFCIs protect the people using devices connected to the circuit. GFCI circuitry may be incorporated into receptacles for localized protection, into breakers that protect an entire branch circuit, or in power cords such as in hair dryers (Figure 5.17). The investigator should not be misled in thinking that the presence of a GFCI will protect against all short circuits. Contact between the energized conductor and the neutral can still lead to arcing/sparking. The GFCI protects against contact between the energized conductor and any *grounded* object.

**Figure 5.17** A typical GFCI wall outlet.

## Electrical Safety Issues
### How Could the System Remain Energized After the Fire?

An electrical system is normally de-energized during fire suppression activities. This is generally accomplished by the local electric utility removing the meter on a single-family dwelling or opening the disconnects at larger facilities. However, building wiring systems can still remain energized by customers bypassing metering equipment (theft), customers having extension cords run from adjacent buildings (after being disconnected for nonpayment of service), or the presence of standby or uninterruptible power supplies for computer or communication equipment.

In addition, if the investigation activities are taking place even as soon as a few hours after the fire, temporary service may have been restored to some circuits.

### Check to Be Sure!

It is not necessary to use an expensive meter to determine whether the electrical system is energized. Two types of devices are designed to give just a simple indication — energized or not. The simpler and less expensive is shown in Figure 5.18. It consists of two leads and a neon light that visually indicate the presence of any voltage from 120 to 600 volts. The advantage of this device is that no batteries are required for it to operate. The disadvantage is that it requires a direct connection from the line being tested to neutral or ground, and the light is quite difficult to see in direct sunlight.

The second type of device is completely insulated and provides a visual and/or audio indication of the presence of voltage by merely holding the device near the suspected energized components. No direct contact is necessary, but the device requires batteries to operate.

### Ground if Necessary!

If you are concerned that someone might reenergize all or some of the system you are examining, you can "ground" the appropriate circuits. This involves actually connecting the involved circuits (or appropriate parts of the panelboard) to ground with a temporary conductor. An electrician should ground the appropriate circuits for this operation to be done safely and properly.

### Use of an Ohmmeter

There are three basic types of electrical meters (often combined into one actual device): voltmeter (to measure voltage), ammeter (to measure current), and ohmmeter (to measure resistance). Because few fire investigators will be measuring voltages or currents on energized circuits, instructions on the use of these functions are left to the manuals provided with the meter.

Investigators often use the ohmmeter, and the operation and use are discussed here. The investigator will often use the *ohmmeter* to trace branch circuits, test fuses and circuit breakers, and check thermostats and other contacts in a circuit. The meter works as shown in Figure 5.19. The device will normally apply a low voltage to the circuit or component being tested. If there is a path for current to flow, the meter on the device will provide an indication. As in the case of voltmeters and ammeters, the manufacturer's instruction manual should be studied thoroughly before the instrument is used. Some important points to consider when using an ohmmeter are:

- An ohmmeter is **NEVER** used on any circuit or device that might be energized. If, for example, an ohmmeter's leads are inserted into a receptacle to check for a short in the wiring, and the branch circuit is energized at its normal 120 volts, the ohmmeter will be damaged and the user could be injured.

- A fire coats wiring, etc., with various forms of oxidation and combustion products. These will need to be removed from the conductor in order for an accurate reading to be made.

- The contacts of switches, thermostats, etc., may also be coated as a result of the fire and may measure as open even though they are actually closed. Experienced electrical investigators using a device called a megohmmeter (Megger™ is the trademark name of one of these devices) can obtain accurate measurements of these types of devices exposed to fire products. These devices supply a high voltage to the circuit being tested; therefore, only experienced personnel should use them. Precautions must be taken to protect any personnel who might be working elsewhere on or near the circuit.

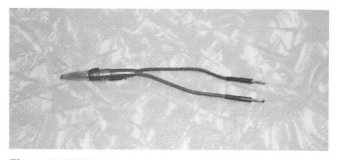

**Figure 5.18** This simple device can be used to test for power in a circuit.

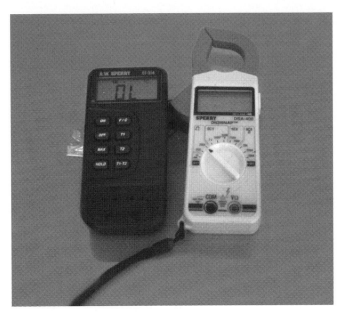

**Figure 5.19** Typical Ohmmeter.

## ◆ Conclusion

This chapter is intended to provide the fire investigator with a basic overview of electricity and the components of a residential electrical system. Electrical systems that are installed to code and function properly are normally very safe. Should the investigator suspect that the source of ignition was the electrical system, then the condition that resulted in the failure should be identified.

Should the investigators have to deal with more complex three-phase electrical systems found in commercial, industrial, or other large facilities, they should seek the assistance of an investigator or electrical expert with experience in those systems. These systems are extremely complex and are beyond the scope of this manual.

# Chapter 6

# The Investigator's Role in Recognizing Hazardous Materials

The fire investigator will rarely be the first responder to a hazardous materials incident. However, the investigator is regularly one of the last remaining fire department personnel at the scene. The fact that investigators work alone or in small numbers at fire scenes that may contain hazardous materials makes the recognition of those materials a crucial requirement of the job. The focus of this chapter is to give the investigator the knowledge needed to identify potentially dangerous materials or conditions found at a fire scene. This chapter does not address the handling of the hazardous materials or incident management but deals only with recognition as it relates to investigator safety. If potentially hazardous materials are found in the course of an investigation, the investigator should minimize his exposure and make the notifications necessary to resolve the problem.

Throughout the chapter, references are made to NFPA 472, *Standard on Professional Competence of Responders to Hazardous Materials Incidents*

(1997). While the entire standard is not of interest for our purposes, we will focus on sections which deal with hazard recognition.

One of the keys to investigator safety is the ability to recognize the presence of hazardous materials. To properly complete this task, a definition of hazardous materials or dangerous goods must be established. Every department, state, province, and governmental organization will have its own definition of a hazardous material. The following definitions for hazardous materials and dangerous goods are provided for the reader's information and knowledge:

- *Hazardous materials — As defined by the U.S. Department of Transportation [DOT] in 49 CFR 171.8 —* A substance or material, including a hazardous substance, that has been determined by the Secretary of Transportation to be capable of posing an unreasonable risk to health, safety, and property when transported in commerce, and which has been so designated.

- *Hazardous materials* — *As defined by NFPA 472* — A substance (solid, liquid, or gas) that when released is capable of creating harm to people, the environment, and property.

- *Dangerous goods* — *As defined by the Canadian Transportation Commission* — Any product, substance, or organism included by its nature or by the regulation in any of the classes listed in the schedule. (**NOTE**: The schedule is the nine United Nations Classes of Hazardous Materials.)

For the remainder of this manual, we will use the term *hazardous materials* and its definition as given in NFPA 472. This definition should enable the investigator to differentiate between a hazardous materials incident and other emergencies.

A *hazardous materials incident* is one that involves a substance that has been released, is threatened by fire, or is on fire. Because of this, the material poses an unreasonable risk to people, the environment, and property. It is almost certain that hazardous materials incidents will be more complex than most other incidents to which the fire service is expected to respond.

Should investigators locate materials or containers while working at a fire scene, they should leave the area until the material is identified and isolated, or removed if necessary. In cases where an investigation must be conducted in an area that cannot be made safe, properly trained investigators may have to enter the area using the appropriate protective equipment.

 **Methods of Identification**

Having an understanding and a workable definition in hand, the development of skills to identify hazards is the next step. There are both formal and informal methods of identifying hazardous materials. The informal ways are generally related to circumstantial evidence at the scene and are looked upon as clues rather than hard facts. Examples of informal clues are reports from bystanders or responsible parties at the scene. By observing indicators, such as smoke color, vapor clouds, or pressure-relief valves on tanks, the investigator is able to begin hazard assessment. The use of personal senses places the investigator in a potentially

dangerous position before identification is confirmed. For this reason, the use of informal methods should never be the sole indicator for hazard identification.

A more reliable approach is positive identification. Only after confirmed identification can a course of action be determined. By surveying the hazardous materials incident from a safe location, recognition can be accomplished through the use of formal identification methods. By identifying the hazard before entry, the amount of risk can be reduced. When a hazardous material is present, the investigator should always wait until the material has been identified and isolated before entering. Knowledge of the following will assist the investigator in formally identifying hazards prior to entry:

- Occupancy and location
- Type and container shape
- Placards and labels
- Shipping papers
- Monitoring and detecting devices
- MSDS information

## Occupancy and Location

Before entering the scene, investigators should educate themselves about the occupancy and location of the incident. Investigators are not under the same time constraints as the first responders and should take time to familiarize themselves with the scene prior to entering. The investigator should determine the following information:

- Type of business
- Product manufactured
- Supplies on hand
- Storage facilities
- Chemicals normally associated with this industry
- Exposures

This information can be gained through prefire planning, previous building inspections, and community involvement. Knowledge of potential hazards in a building or investigation site is a personal safety issue for the investigator. In general, the emergency nature of an incident has passed when the investigator begins to work. The scene needs to be free of any related hazards before investigators or other nonsuppression personnel are allowed to enter.

## Type and Shape of Container

The type and shape of a container can be a valuable indicator for recognizing the substance involved. There are two basic categories of storage containers: fixed and mobile. Within these two categories are many subcategories. Figures 6.1 a through d show examples of fixed, mobile, and rail containers. The investigator should be able to visually identify all types from as far away as is appropriate.

While pre-incident planning is looked upon as one of the best sources of hazard recognition, the investigator will not always have this luxury. The investigator can encounter substances that are stored or being transported in unexpected locations. The difficulties in recognizing these substances will range from no markings or mismarking to apathy or ignorance of the regulations. Two resources that are available to assist in the identification of materials found in the field are the 1996 *North American Emergency Response Guidebook* and the NFPA 704, *Standard System for the Identification of the Hazards of Materials for Emergency Response* marking system for fixed locations.

In the early 70s, the Department of Transportation (DOT) recognized a need to aid first responders in dealing with hazardous materials incidents. At that time, it identified a total of eleven hazardous materials and the appropriate responses for each. The guide has grown in both size and importance as the years have progressed — from the second edition in 1974 that had 31 identifiable hazards to the most recent 1996 edition that has roughly 3,000 identified hazards.

Before 1996 there were several resources that overlapped one another to provide information for first responders in North America. In 1996, there was unification for an emergency response guide throughout North America. The 1996 *North American Emergency Response Guidebook* (NAERG96) was developed by Transport Canada, the U.S. Department of Transportation, and the Secretariat of Communications and Transportation for Mexico. The guide was developed for use by first responders such as firefighters, law enforcement, and emergency medical services personnel. A comprehensive understanding of how to use the guide is a tool that the investigator can call upon to aid in hazard recognition. The book is the same format as past ERGS but caters to a wider spectrum by encompassing Canada and Mexico.

The NAERG96 is a basic guide for first responders to use during their initial actions to protect

**Figure 6.1a** Tankers carry a wide variety of chemicals.

**Figure 6.1b** Some railroad tank cars are sole-use carriers.

**Figure 6.1c** Compressed gas trailers are obvious by their appearance.

**Figure 6.1d** A typical fixed chemical tank.

themselves and the general public while intervening in an incident. The guide provides the user with basic information that can be used to identify the existence of potential hazards. The NAERG is an excellent tool for emergency responders and related personnel, such as fire investigators, who respond to and work at emergency scenes.

The guide is user-friendly, taking the reader step by step through necessary information and precautions. The sections are color-coded, which allow the reader to take whatever information is available and cross-reference it.

## United Nations Classification System

Both the United States and Canada have adopted the United Nations (UN) system for classifying and identifying hazardous materials transported both internationally and domestically. Under this system, nine hazard classes are used to categorize hazardous materials. In addition to these nine classes, a separate category exists for other regulated materials (ORM-D). The nine hazard classes used for categorizing hazardous materials are as follows:

- Class 1 - Explosives
- Class 2 - Gases
- Class 3 - Flammable Liquids
- Class 4 - Flammable Solids
- Class 5 - Oxidizers
- Class 6 - Poisons and Infectious Substances
- Class 7 - Radioactive Substances
- Class 8 - Corrosives
- Class 9 - Miscellaneous

The UN system forms the basis for the DOT regulations. The DOT classifies hazardous materials according to their primary danger and assigns standardized symbols to identify the classes. This is similar to what the UN system has done. DOT regulations cover several other types of substances in addition to the nine classes identified in the UN system. The major classes and a brief description of each are given in Table 6.1.

Hazardous materials that are not transported but only used, produced, or stored at a fixed site do not fall under the same regulations as transported materials. A very effective tool for fixed sites is the system defined by NFPA 704, *Standard System for the Identification of the Hazards of Materials for Emergency Response* (1996). This is useful in both pre-incident planning and scene investigation. The investigator should have the ability to look at the NFPA 704 placard and determine what risks are being undertaken.

Specifically, the NFPA 704 system uses a rating system of zero (0) to four (4). A zero indicates there is no hazard present, and a four represents a severe hazard. The rating is assigned to three categories: health, flammability, and reactivity. The rating numbers are arranged on a diamond-shaped marker or sign. The health rating is located on a blue background at the nine o'clock position. The flammability hazard rating is positioned on a red background at the twelve o'clock position. The reactivity hazard rating appears on a yellow background and is positioned at three o'clock. As an alternative, the backgrounds for each of these rating positions may be any contrasting color, and the numbers (0 to 4) may be represented by the appropriate color (blue, red, and yellow). Special hazards are located in the six o'clock position and have no specified background color; however, white is most commonly used.

The ratings for each hazard (health, flammability, and reactivity) are described in Table 6.2. This table also describes the special hazards that may be indicated on the NFPA 704 marker. The NFPA 704 system is used in conjunction with NFPA 49, *Hazardous Chemicals Data*. NFPA 49 describes the properties and hazards of various materials and provides information on personal protection and fire fighting when facing these specific chemicals. Valuable information is given on assigning appropriate ratings to the NFPA 704 markers at facilities that contain listed chemicals.

## Material Safety Data Sheet (MSDS)

The best source of information on a specific hazardous material is the manufacturer's data sheet known as a material safety data sheet (MSDS). State and federal legislation on hazard communication, right-to-know, and mandatory local notification on hazards make the MSDS a necessity. Fire investigators can acquire an MSDS from the manufac-

## Table 6.1

| Hazard Class | Product Example |
|---|---|
| **1 Explosives** | |
| 1.1 Mass explosion hazard | Black powder |
| 1.2 Projection hazard | Detonating cord |
| 1.3 Fire hazard | Propellant explosives |
| 1.4 No significant blast | Practice ammunition |
| 1.5 Very sensitive | Prilled ammonium nitrate |
| 1.6 Extremely insensitive | Fertilizer fuel-oil mixtures |
| **2 Gases** | |
| 2.1 Flammable gas | Hydrogen |
| 2.2 Nonflammable gas | Nitrogen |
| 2.3 Poisonous gas | Phosgene |
| **3 Flammable Liquids** | Gasoline, kerosene, diesel fuel |
| **4 Flammable Solids, Spontaneously Combustible Materials, and Materials that are Dangerous When Wet** | |
| 4.1 Flammable solids | Magnesium |
| 4.2 Spontaneously combustible | Phosphorus |
| 4.3 Dangerous when wet | Calcium carbide |
| **5 Oxidizers and Organic Peroxides** | |
| 5.1 Oxidizers | Ammonium nitrate |
| 5.2 Organic peroxides | Ethyl ketone peroxide |
| **6 Poisonous and Etiologic Materials** | |
| 6.1 Poisonous | Arsenic |
| 6.2 Infectious (etiological agent) | Rabies, HIV, Hepatitis B |
| **7 Radioactive Materials** | Cobalt |
| **8 Corrosives** | Sulfuric acid |
| **9 Miscellaneous Hazardous Materials** | |
| 9.1 Miscellaneous (Canada only) | PCBs, molten sulfur |
| 9.2 Environmental Hazard (Canada only) | PCB, asbestos |
| 9.3 Dangerous Waste (Canada only) | Fumaric acid |
| **ORM-D (Other regulated materials)** | Consumer commodities |

turer of the material, the supplier, the facility hazard communication plan, or the local emergency planning committee.

Minimal content of the MSDS is mandated by the U.S. Department of Labor, Occupational Safety and Health Administration (OSHA). While the actual format of the information may vary, each sheet will have eight sections that contain the following information:

*Section I*
- Manufacturer's name and address
- Emergency telephone number
- Information telephone number
- Signature and date

*Section II - Hazardous Ingredients*
- Common name
- Chemical name
- CAS number
- OSHA Permissible Exposure Limit (PEL)
- ACGIH Threshold Limit Value (TLV)

## Table 6.2
## NFPA 704 Rating System

| Identification of Health Hazard | | Identification of Flammability | | Identification of Reactivity | |
|---|---|---|---|---|---|
| Type of Possible Injury | | Susceptibility of Materials to Burning | | Susceptibility to Release of Energy | |
| Signal | | Signal | | Signal | |
| 4 | Materials that on very short exposure could cause death or major residual injury. | 4 | Materials that will rapidly or completely vaporize at atmospheric pressure and normal ambient temperature, or that are readily dispersed in air and that will burn readily. | 4 | Materials that in themselves are readily capable of detonation or of explosive decomposition or reaction at normal temperatures and pressures. |
| 3 | Materials that on short exposure could cause serious temporary or residual injury. | 3 | Liquids and solids that can be ignited under almost all ambient temperature conditions. | 3 | Materials that in themselves are capable of detonation or explosive decomposition or reaction but require a strong initiating source or which must be heated under confinement before initiation or which react explosively with water. |
| 2 | Materials that on intense or continued but not chronic exposure could cause temporary incapacitation or possible residual injury. | 2 | Materials that must be moderately heated or exposed to relatively high ambient temperatures before ignition can occur. | 2 | Materials that readily undergo violent chemical change at elevated temperatures and pressures or which react violently with water or which may form explosive mixtures with water. |
| 1 | Materials that on exposure would cause irritation but only minor residual injury. | 1 | Materials that must be preheated before ignition can occur. | 1 | Materials that in themselves are normally stable, but which can become unstable at elevated temperatures and pressures. |
| 0 | Materials that on exposure under fire conditions would offer no hazard beyond that of ordinary combustible material. | 0 | Materials that will not burn. | 0 | Materials that in themselves are normally stable, even under fire exposure conditions, and which are not reactive with water. |

- Other exposure limits

*Section III - Physical and Chemical Characteristics*
- Boiling point
- Specific gravity
- Vapor pressure
- Melting point
- Vapor density
- Evaporation rate

- Solubility in water
- Appearance and odor

*Section IV - Fire and Explosion Hazard Data*
- Flash point
- Flammable limits (LEL, UEL)
- Special fire fighting procedures
- Unusual fire and explosion hazards

*Section V - Reactivity Data*
- Stability (stable/unstable conditions to avoid)

- Incompatibility (materials to avoid)
- Hazardous decomposition or by-products
- Hazardous polymerization (may or may not occur, conditions to avoid)

*Section VI- Health Hazard Data*
- Routes of entry
- Health hazards (acute or chronic)
- Carcinogenicity
- NTP (National Toxicological Program)
- IARC (International Agency for Research on Cancer) monographs
- OSHA regulated
- Signs and symptoms of exposure
- Medical conditions aggravated by exposure
- Emergency and first-aid procedures

*Section VII - Precautions for Safe Handling and Use*
- Waste disposal methods
- Handling and storing precautions
- Other precautions

*Section VIII - Control Measures*
- Respiratory protection
- Ventilation (local, mechanical, special, other)
- Protective gloves
- Eye protection
- Other protective clothing or equipment

The MSDS should be looked upon as a tool that can aid the investigator in identifying hazards. Investigators should be familiar with the eight sections and be able to recognize the areas that indicate the presence of hazardous materials.

Another place that the MSDS can be found is with the shipping papers. Most shipments of hazardous materials must be accompanied by shipping papers that describe the hazardous material. For most shipments, DOT does not specify the use of a particular type of document. The information can be provided on a bill of lading, waybill, or similar document (Figure 6.2). The exceptions are hazardous waste shipments, which must be accompanied by a document called a *Uniform Hazardous Waste Manifest*. Instructions for describing hazardous materials are provided in the Department of Transportation/Transport Canada regulations. These descriptions include the following:

- Proper shipping name of the material
- Hazard class represented by the material

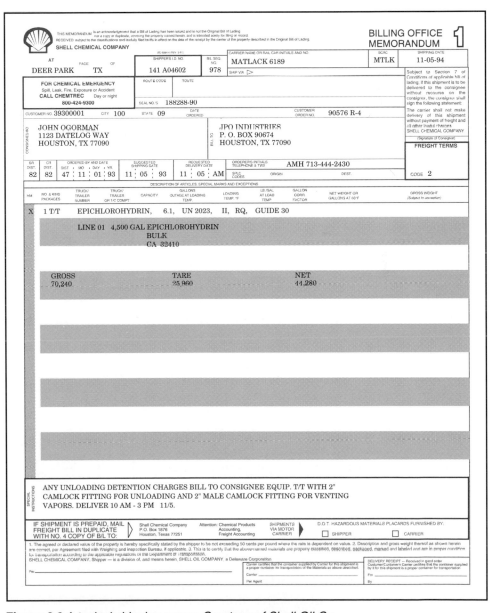

**Figure 6.2** A typical shipping paper. *Courtesy of Shell Oil Co.*

| Table 6.3 Shipping Paper Identification | | | |
|---|---|---|---|
| **Transportation Mode** | **Shipping Paper Name** | **Location of Papers** | **Party Responsible** |
| Air | Air bill | Cockpit | Pilot |
| Highway | Bill of lading | Cab of vehicle | Driver |
| Rail | Waybill/consist | Engine or caboose | Conductor |
| Water | Dangerous cargo manifest | Bridge or pilothouse | Captain or master |

- Packing group assigned to the material
- Quantity of material

In addition, special description requirements apply to certain types of materials (for example, those that cause poison by inhalation, radioactive materials, and hazardous substances) and modes of transportation.

Once the fire investigator has determined that a close approach to an incident is safe, he should then examine the cargo shipping papers. The location and type of paperwork change according to the mode of transport (Table 6.3). In each of the following modes, the general area is given. However, the exact location of the documents will vary. Investigators may need to check the entire general area in order to locate these documents. In trucks and airplanes, these papers are placed near the driver or pilot. On ships and barges, the papers are placed on the bridge or in the pilothouse of a controlling tugboat. On trains, the waybills and consist may be placed in the engine, caboose, or both. During pre-incident planning, the location of the papers for a specific rail line can be determined.

Having a complete and thorough understanding of where to find and how to use the shipping papers and material safety data sheets is invaluable information for the investigator. These simple but effective tools will allow the investigator to positively identify a hazardous material. Once identification is confirmed, the investigators are able to prevent any unnecessary harm to themselves or others.

## Hazard Recognition
The investigator's ability to grasp the use of the tools at his disposal is an important component of the job. While the use of specific tools such as the NAERG96, NFPA 704, and MSDS has been

covered, there are certain hazards that fall between the gaps of these resources. These hazards include but are not limited to the following: military markings, pesticide labels, pipeline identification, and PCB identification (polychlorinated biphenyl's).

The United States military and the Canadian military have established their own marking systems for hazardous materials and chemicals. The key component the investigator needs to consider when confronted with this type of situation is to remember that military ordnance is designed to inflict bodily harm and/or heavy property damage. Recognition of the following symbols and classes

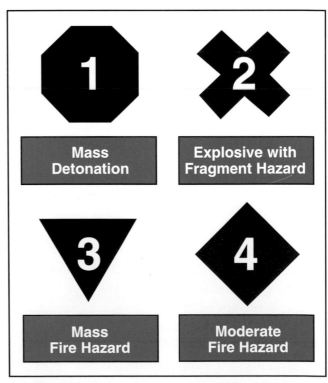

**Figure 6.3a** Military markings for fire and explosion hazards.

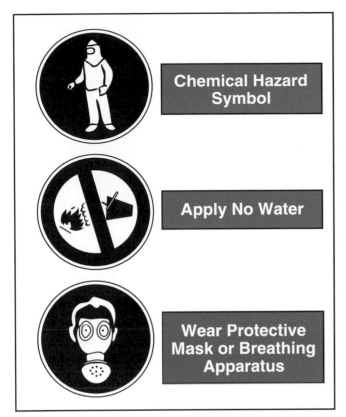

Figure 6.3b Military markings for chemical hazards.

Figure 6.4 Typical LPG pipeline markers.

will aid the investigator in confirming a military hazard (Figures 6.3 a and b).

Fire investigators should develop a general familiarity with pesticides and their labeling. The EPA regulates the manufacturing and labeling of all pesticides. Each EPA label must contain one of the following signal words:

- **Danger/Poison** — denotes highly toxic materials

- **Warning** — denotes moderate toxicity

- **Caution** — denotes relatively low toxicity

- **Extremely Flammable** — denotes flashpoint below 80 degrees

Many types of materials, particularly petroleum products, are transported across the country in an extensive network of pipelines. The investigator need only be aware of the different types of markers and the hazards they represent (Figure 6.4).

Additional hazards that may be encountered are containers, transformers, or capacitors that contain polychlorinated biphenyl's (PCBs). The investigator can use EPA labels and DOT labels to identify and confirm the presence of PCBs.

## ◆ Conclusion

Today's fire investigators are being exposed to potentially hazardous conditions and materials on a regular basis. The basic skills and tools that have been presented in this chapter should give the investigator a solid hazardous materials foundation on which to build. While the identification of hazardous materials is inherently difficult, this difficulty can be overcome through continued study and education. The ability to recognize and identify potentially hazardous materials encountered on the job will make the fire investigator's job a safer one.

# Chapter 7

# Examining the Fire Scene

The previous chapters of this manual presented the basic knowledge that the fire investigator needs to conduct a fire investigation. This chapter deals with the collection of information at the fire scene that the investigator uses to develop an opinion on the origin, cause, and responsibility for the fire.

A fire investigation is very much like solving a puzzle. Fire investigators must develop information about the fire and fit these pieces of information together in order to form their opinions. These individual pieces of information may be in the form of statements made by witnesses or involved parties, physical evidence collected at the fire scene, or information developed through research after the fire. The fire investigator's opinions are then based on an analysis of all information collected.

 **The Investigative Process**

The investigation almost always begins with an examination of the fire scene. The proper examination of both the exterior and interior of structures involved in the fire is a critical component of the investigation. The scene is the place where much of the evidence related to the fire will be found. The scene and related evidence may be very fragile as a result of the fire. Witnesses present at the scene may disperse after the incident and be very difficult to relocate. Access to the building may be restricted after the initial phases of the fire. For all these reasons, the initial processing of the fire scene should take place as soon as possible after the fire. Investigators should keep in mind that they may have only one opportunity to view and document the scene, collect the physical evidence available, and conduct interviews that will be critical to the development and support of their opinions.

The process flowchart presented in Figure 7.1 is intended to provide the investigator with a visual representation of the steps that must be completed during an investigation. The reader should keep in mind that almost every incident is different and that the investigation may have to be accomplished in a different order than presented in this chart. This chapter discusses each of the components of the process related to processing the fire scene. The flowchart in Figure 7.1 also correlates the applicable JPRs found in NFPA 1033, *Standard for Professional Qualifications for Fire Investigator* (1998), to the specific stages of the investigation.

It is important for the fire investigator to develop a methodology or system that will be used to conduct investigations. NFPA 921, *Guide for Fire and Explosion Investigations*, suggests using a systematic approach, such as the scientific method explained in the following sidebar. While this may work for some, others may decide on a different model. No matter what method is used, investigators must find a reliable method that works and use it consistently. Fire investigators must be able to defend their investigative methodology whenever they present their findings in court.

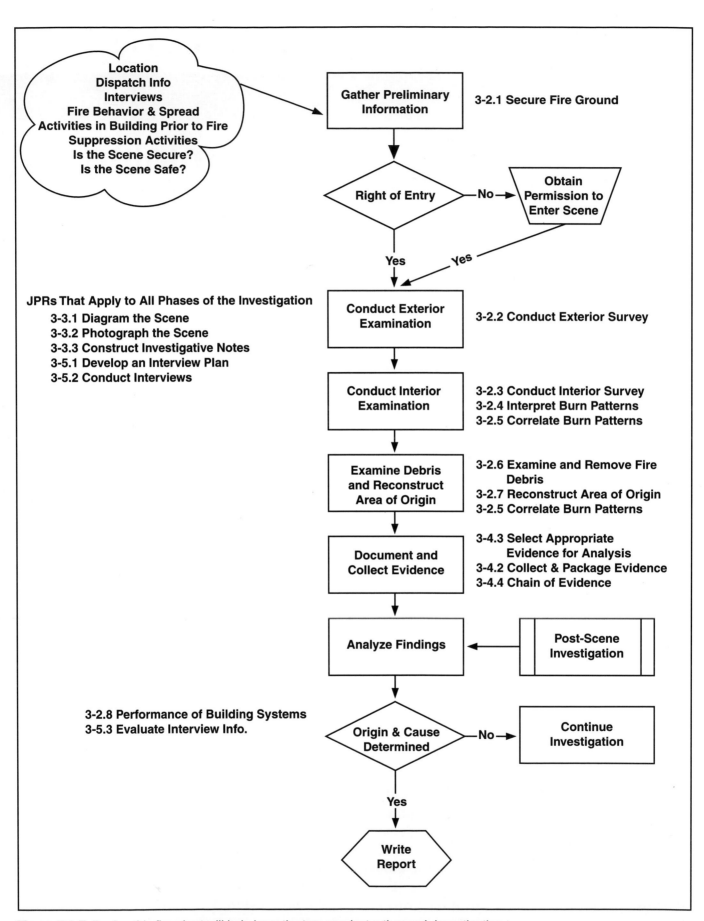

**Figure 7.1** Following this flowchart will help investigators conduct a thorough investigation.

### What is a Systematic Approach?

In most fire investigations, the sequence of the investigation will be to first determine the area or point of origin — where the fire began. When that is accomplished, the investigator then begins to attempt to determine the material that was first ignited (fuel) and the source of the heat that caused the ignition. The final determination would be to identify the factors (ignition factors) that brought the heat and fuel together in such a way that an ignition resulted. The process flowchart shown in Figure 7.1 provides an example of a systematic approach to an investigation. One systematic approach for the conduct of fire investigations is the use of the scientific method, as suggested by NFPA 921. For a fire investigation, the scientific method would be applied as follows:

- **Recognize the need** — The fact that a fire or explosion has occurred and you have been assigned to investigate shows the need.

- **Define the problem** — Once an incident has occurred, then an investigation into the origin, cause, and responsibility should take place. All the steps shown in Figure 7.1 will be followed as part of the investigation.

- **Collect data** — During the investigation, the facts related to the incident are collected. The data may be physical evidence gathered at the fire scene, information developed based on the testing of materials collected at the scene, or the observations of the investigator while conducting the investigation.

- **Analyze the data** — The information collected during the investigation is then analyzed based on the knowledge, training, and experience of the fire investigator. During this phase of the investigation, only that information which can be substantiated by observation or experiment should be included in the analysis.

- **Develop a hypothesis** — At this point, the investigator uses the facts that are available to develop an assumption or set of assumptions regarding the origin and cause of the fire. This set of assumptions (hypothesis) should be based only on the facts currently available to the investigator.

- **Test the hypothesis** — Test the validity of the assumptions made by eliminating all other reasonable origins and causes. Assumptions are compared with known facts developed during the investigation. If the assumptions about the origin and cause cannot stand up to this test, the fire investigator should discard them and develop a new set of assumptions. To accomplish this, existing information is reexamined and if necessary, additional information (data) collected. When all the available data is collected and analyzed and an expectable set of assumptions cannot be developed, the cause of the fire is "undetermined."

The process is simple and remains the same for every investigation. For incidents where the point of origin is rapidly identified and the source of the heat of ignition can be determined — and other potential sources ruled out — following this method may take only a few minutes. Where a building is severely damaged and there are numerous points where low burning occurred, the process of identifying the point or points of origin and the related elimination of heat sources may be very time consuming and require the development and testing of numerous sets of assumptions that must be "tested" before a determination can be made — if one can be made at all.

 ## Preliminary Information and Considerations

Each investigator assigned to a fire will have to develop or review some preliminary information regarding the incident. This information may be the dispatch information received by the first investigator assigned to the fire or the case information provided to an insurance investigator assigned later in the investigation. Investigators assigned to a fire will have to determine the location of the incident. They may also collect the time and date of the fire and the type of occupancy of the fire building. On arrival at the fire scene, the investigators should also evaluate the security of the scene and determine whether it is safe for them to enter and conduct their examination. If security or scene safety is not satisfactory, it should be addressed immediately. The investigator must also ensure that he has the right to enter the scene. Once these areas meet the investigator's satisfaction, the scene examination can be conducted.

### Initial Interviews

If fire suppression operations are still ongoing when the investigator arrives, initial interviews of firefighters and officers involved in the initial fire attack may be conducted. The investigator may also interview any available eyewitnesses to the ignition or early stages of the fire along with those with information related to the fire scene (Figure 7.2). These interviews should be aimed at determining the following information:

- Location, size, behavior, and spread of the fire.

- Activities in or around the building before or at the time the fire was discovered.

- Unusual circumstances or activities in or around the building before the discovery of the fire, such as domestic disputes, strangers in the area, lack or change of normal activity, behavior of occupants, strange sounds, electricity loss to structures, flickering lights, and circuit breakers tripping.

- Environmental factors such as weather, earthquakes, floods, etc.

- Information regarding fire department suppression activities. Was the building secured on its arrival?

- Building construction, structural layout, condition, fuel load, fuel arrangement, building systems.

The timing of these interviews is up to the investigator. Some investigators may want to conduct a preliminary examination of the scene before interviewing anyone. Others may decide to do the interviews before the scene examination. The investigator should consider timing the initial in-

terviews based on the availability of witnesses and the condition of the scene. Regardless of when the interviews are conducted, the fire investigator should use proper techniques and follow established organizational procedures. See Chapter 11, "Interviewing and Interrogation," for specific information on conducting interviews and the types of questions that should be asked during these interviews. The proper documentation of all interviews conducted is a critical step in the investigation (Figure 7.3). At a minimum, the investigator should document the following:

- Name of the individuals interviewed

- Where individuals can be contacted should there be additional questions

- An overview of statements made during each of the interviews

## Right of Entry

Another important consideration in the early stages of the scene examination is whether or not the investigator has the legal right to enter the fire building and collect potential evidence related to

**Figure 7.2** Witnesses should be interviewed as soon as possible. *Courtesy of Sheldon Levi.*

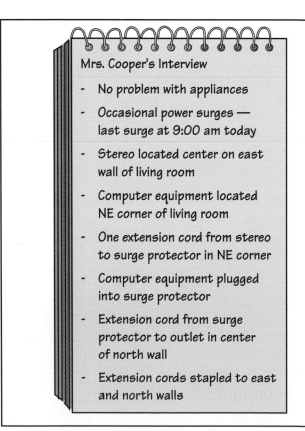

Mrs. Cooper's Interview

- No problem with appliances

- Occasional power surges — last surge at 9:00 am today

- Stereo located center on east wall of living room

- Computer equipment located NE corner of living room

- One extension cord from stereo to surge protector in NE corner

- Computer equipment plugged into surge protector

- Extension cord from surge protector to outlet in center of north wall

- Extension cords stapled to east and north walls

**Figure 7.3** Interviews should be thoroughly documented.

the fire. The fire investigator may obtain the right to enter a scene through several means or methods including:

- Consent
- Exigent circumstances
- Criminal search warrant
- Administrative search warrant (Some jurisdictions do not provide the means to obtain administrative warrants)
- Court order

(**NOTE:** For more information and explanation, see the sidebar on page 100.)

If firefighters maintain possession of the premises, for a reasonable time, to prevent unauthorized entry and to ensure that the area is not disturbed until the investigator arrives, the investigator may not need to obtain a search warrant to conduct the scene examination (Figure 7.4). (See Appendix B for an explanation of reasonable time.) This point concerning a search warrant is based on *Michigan vs. Tyler* (436 U.S. 499, 56 L.Ed. 2d 486 [1978]). The U.S. Supreme Court held in that case that "once in a building [to extinguish a fire], firefighters may seize [without a warrant] evidence of arson that is in plain view .... [and] officials need no warrant to remain in a building for a reasonable time to investigate the cause of a blaze after it has been extinguished."

The U.S. Supreme Court agreed, with modification, with the Michigan State Supreme Court's statement that "[if] there has been a fire, the blaze extinguished, and the firefighters have left the premises, a warrant is required to reenter and search the premises, unless there is consent . . . ." (See Appendix B for more discussion on the *Michigan vs Tyler* decision.)

Another case, Michigan vs Clifford (Appendix B), further stressed the importance of following proper search warrant procedures to ensure the admissibility of evidence in a court of law. In the "Clifford" case, the fire department had left the premises and the property owners had arranged for securing the structure. Entry by investigators, approximately five hours after the fire department departure, to determine origin and cause of the fire was made without consent, an administrative search warrant, or exigent circumstances. Evidence obtained inside the structure due to this improper search was found to be inadmissible. The court decision was that an administrative search warrant was sufficient but necessary to reenter the structure to determine origin and cause of the fire. The court further found that once the cause was determined to be arson, any additional search beyond the area of origin would require a criminal search warrant.

In the "Clifford" case, the area of origin was discovered to be the basement, and evidence found elsewhere in the structure was deemed to be beyond the scope of determining the cause and origin of the fire. This was due to the fact that a criminal search warrant was not obtained and that once the cause of arson had been determined, any additional search would have been for further evidence of the crime.

The impact of these decisions seems to be that if there is evidence of possible arson, the fire department should leave an engine company or a law enforcement officer on the premises for a reasonable time until the investigator arrives. What constitutes a reasonable time is based on policy, the circumstances surrounding the incident, and case law. To leave the premises, return later without a search warrant, and then make a search might be sufficient grounds to make prosecution impossible or for an appellate court to overturn a conviction.

Fire investigators must have a thorough knowledge of the policies and legal opinions that affect their jurisdiction in this regard. These opinions or interpretations can be obtained from such persons as the district attorney or state attorney general. A standard operating procedure should be written around these policies and opinions.

**Figure 7.4** Scene control should be maintained as long as reasonably necessary.

## Search Warrants

A warrant is not required by the fire department to enter property to suppress fire; this is called an "exigent circumstance." *Exigent circumstances* are defined by Blacks Law Dictionary as "situations that demand unusual or immediate action." Firefighters may also remain on the scene for a reasonable amount of time to conduct salvage and overhaul and origin and cause determination. Once the fire scene has been released, investigators may not reenter the property without first gaining the consent of the owner or obtaining a warrant. There are two types of warrants for entry into a structure: administrative and criminal.

### Administrative

An administrative warrant is required in most jurisdictions in order to obtain entry after the fire department has left the scene. The following is quoted from the Michigan vs. Tyler case concerning administrative warrants:

"To secure a warrant to investigate the fire, an official must show more than the bare fact that the fire has occurred. The magistrate's duty is to ensure that the proposed search will be reasonable, a determination that requires inquiry into the need for the intrusion on the one hand, and the threat of disruption to the occupant on the other....The number of prior entries, the scope of the search, the time of day when it is proposed to be made, the lapse of time since the fire, the continued use of the building, and the owner's efforts to secure it against intruders might all be relevant factors. Even though a fire victim's privacy must normally yield to the vital social objective of ascertaining the cause of the fire, the magistrate can perform the important function of preventing harassment by keeping that invasion to a minimum."

[Michigan vs. Tyler, 56 L.Ed.2d 486 (1978)]

### Criminal

Once an investigator has probable cause that arson has been committed and additional entries into the scene are required, a criminal search warrant or consent must be obtained before a search for evidence is conducted.

In summary of search warrants, the Michigan vs Tyler case states the following:

"We hold that an entry to fight a fire requires no warrant, and that once in the building, officials may remain there for a reasonable amount of time to investigate the cause of the blaze. Thereafter, additional entries to investigate the cause of the fire must be made pursuant to the warrant procedures governing administrative searches.

Evidence of arson discovered in the course of such investigations is admissible at trial, but if the investigating officials find probable cause to believe that arson has occurred and require further access to gather evidence for a possible prosecution, they may obtain a warrant only upon a traditional showing of probable cause applicable to searches for evidence of crime."

*[Michigan v. Tyler 56 L.Ed.2d 486 (1978)]*

Use the following guidelines regarding warrants:

- Entry by fire department to fight fire does not require a search warrant.
- Fire department can stay on the fire scene for a reasonable period of time to determine origin and cause of fire.
- Any evidence found in plain view during the suppression or origin and cause determination usually is admissible in court.
- If the investigators arrive after the fire scene has been released, consent or a warrant must be obtained to reenter.
- If during a scene examination (conducted with an administrative warrant) the investigator determines the cause of the fire to be arson, the investigator should stop the investigation and obtain a criminal search warrant or the consent of the owner.
- If the cause is determined to be arson and search for further evidence is needed, consent or a criminal search warrant must be obtained for each reentry.

Private investigators representing insurance companies or other private interests must operate under the jurisdiction of the organization by which they are employed. Most insurance policies contain language that allows the insurer to investigate insured losses. In spite of this, the insured has the right not to allow the investigator on the scene. Therefore, the fire investigator should ensure that permission to be on the premises to conduct an investigation or to take evidence has been obtained through court order or by verbal or written consent. In most cases, this permission is issued one time for the number of visits necessary to gather the required information for the client. In the case of an insurance claim investigation, the insured has an obligation to cooperate with the insurer in the gathering of facts related to the loss.

Private investigators may be requested to conduct origin and cause investigations at loss sites for entities other than the owner or first party insurance carrier. In these cases the investigator must ensure that appropriate permission or consent has been obtained to enter upon the scene. If any removal of evidence or major alteration of the scene is anticipated, further permission should be obtained.

There are several potential consequences if a private investigator exceeds his legal authority. The case itself could be prejudiced. Evidence collected may be excluded from use in court. Punitive damages or an increased award of cost could be claimed against the investigator's client. Further, the investigator and the person that retained him face potential civil liability. They are exposed to a civil trespass action, or they could be personally named in a bad faith action. Where the private investigator is acting in cooperation with a police or public fire investigator and enters a fire scene without permission or a court order, collusion (civil conspiracy) or malicious prosecution could be alleged.

Once preliminary information has been obtained from those present at the fire scene and the right to legally enter has been determined, the investigator can begin to collect the information needed to determine the origin, cause, and responsibility of the fire. The next seven chapters in this manual deal with the specifics of the scene and its examination.

 ## Conclusion

To determine how a fire started and who was responsible for starting it, fire investigators must use a scientific approach. This means that they must gather information in and around the fire scene, and analyze that information to reach a logical conclusion. In order to gather the necessary information, investigators must have access to the fire scene. This may require that fire suppression personnel remain at the scene until the investigator arrives. Or, if firefighters do not maintain control of the scene, the investigator will have to get the consent of the property owner or obtain an administrative or criminal search warrant to reenter the property.

# Chapter 8

# Securing the Fire Scene

**Performance Objectives**

This chapter provides information that addresses performance objectives described in NFPA 1033, *Standard for Professional Qualifications for Fire Investigator*, particularly those referenced in the following sections:

**Chapter 3 Fire Investigator**
**3-2.1 a and b**

 secure scene is one with a recognizable perimeter and someone to maintain that perimeter. The security of the scene is initially the responsibility of the fire suppression personnel who respond to extinguish the fire. These early measures should include the following:

- Restricting access to the scene

- Protecting any potential evidence located in the area

- Minimizing fire suppression and overhaul activities that could destroy important information regarding the origin and cause of the fire

The first-arriving fire investigator may have to adjust the security measures already in place or implement additional measures to protect the scene. This decision will be based on his assessment of the scene, the circumstances surrounding the fire, and the measures in place.

## ◆ Protecting Evidence

The primary objective for securing the scene is to protect evidence from being contaminated, damaged, or destroyed before it can be documented and preserved or collected. Scene security should also prevent the alteration of the evidence at the scene such as the changing of circuit breaker or appliance controls and/or switch positions. In the legal context, altering, changing, or failing to preserve any potential evidence at the scene is called *spoliation*. By limiting access to the scene to only necessary emergency personnel involved in fire suppression, salvage, overhaul, and the investigation efforts, the scene is protected from excessive foot traffic. Restricting access also reduces the potential possibility that bystanders, building occupants, or individuals involved in the ignition of the fire will enter the scene and remove, contaminate, or destroy evidence (Figure 8.1).

**Figure 8.1** Perimeter control must be maintained to preserve evidence.

### Contamination and Spoliation

Even if a scene is marked off and access is restricted to emergency responders or other authorized personnel, hidden hazards exist that could threaten the integrity of evidence. One such hazard is contamination. A second is spoliation of evidence. Sometimes these hazards arise as unforeseen consequences of routine activities.

### Contamination

*Contamination* is a broad concept, encompassing anything that can taint physical evidence. Examples of potential sources of contamination include:

- Smoking materials, such as cigarette butts or matchsticks, which have been carelessly dropped by firefighters, spectators, or investigators

- Ignitable liquid traces introduced into the scene by items that have been stored near or exposed to ignitable liquids including:
  — Boots/gloves
  — Power cords
  — Tools
  — Power equipment

### Spoliation

*Spoliation* is evidence that is destroyed, damaged, altered, or otherwise not preserved (Figure 8. 2). From the viewpoint of securing the scene, the investigator must recognize and guard against potential spoliation. Spoliation avoidance might require:

- Tarping portions of the scene that are exposed to the weather and that otherwise would be adversely effected from exposure to the elements

- Minimizing destruction during reinforcing or shoring operations

- Minimizing foot traffic on the scene so as not to obscure burn patterns on floor surfaces by tracking over them

A secondary objective for limiting access into the fire scene is safety. As the perimeter is established, known hazardous areas should be marked or otherwise secured to prevent injury to personnel operating in the area (Figure 8.3).

### ◆ Establishing Perimeters

The investigator's first decision on the scene is to determine the size of the perimeter that must be established around the fire scene. The size of the perimeter may change several times during an inci-

**Figure 8.2** Firefighters can destroy evidence if they start the overhaul before getting permission from the investigator. *Courtesy of Joseph J. Marino.*

**Figure 8.3** Hazardous areas should be marked or otherwise secured.

dent based on the situation and progression of the investigation. Guidelines to establishing a perimeter of the proper size are as follows:

- *Explosions.* The perimeter for explosions should be established at 1.5 times the distance from the farthest piece of debris found. As the investigation continues, this perimeter will expand as additional debris is located (Figure 8.4).

- *Building fires.* The initial perimeter may be established to provide firefighters an area in which to work. The investigator may expand the perimeter to encompass the area surrounding the building and any potential evidence that is located outside the building. The perimeter should extend beyond the farthest piece of evidence located during the exterior examination of the structure. If no evidence is found outside the building, it may be sufficient to restrict access into the building or an area of the building. It should provide the investigator with room to work and protect all known evidence without being hampered by unnecessary personnel or bystanders (Figure 8.5).

## Effective Perimeters

To be effective, fire scene perimeters must be both recognizable and enforceable. Common ways to accomplish these ends are as follows:

- Ensure that they are visible and recognizable to everyone on the scene. Methods used to mark the perimeter should be easy to use. To accommodate the rapid establishment of a perimeter that is visible and recognizable, many public safety organizations use rope, traffic cones, or marked barrier tape (Figure 8.6).

- Use uniformed police officers or firefighters to control access into the established perimeter. For long-term operations, consider private security guards and/or the installation of construction barriers or fences (Figure 8.7).

**Figure 8.5** The cordoned area should encompass all evidence.

**Figure 8.6** Fireline tape makes a clearly visible boundary. *Courtesy of Bill Lellis.*

**Figure 8.4** Following an explosion, the perimeter may have to be expanded. *Courtesy of Steve George.*

**Figure 8.7** Firefighters may be used to maintain perimeter control.

If the investigation is considered to be criminal in nature:

- Keep a log of all persons who enter and leave the incident perimeter.

- Permit access only to those individuals who are authorized to be in the area. When firefighters and other emergency personnel have completed their tasks in the area, they should move to a staging area outside the perimeter and wait for additional assignments or release from the incident. Should others be brought into the area, they should always be escorted.

- Mark potential evidence located within the perimeter so that it will not be disturbed before detailed examination, documentation, and collection. Use available materials, such as rope, traffic cones, or barrier tape, to provide this protection.

- Where personnel are still operating and there is a potential for damage due to foot traffic or from the operations being conducted, it may be necessary to station a firefighter or investigator with the evidence to protect it until it can be processed.

 **Conclusion**

The physical security of a fire scene is essential to the proper investigation of a fire incident. This is done to protect the evidence in and around the scene that the fire investigator will use to determine the origin and cause of the fire. Scene security is dynamic, and measures taken will change as information is obtained or evidence is located. Security measures should begin with the arrival of fire suppression forces and end when fire investigators complete the scene examination.

# Documenting the Fire Scene

Documenting the fire scene is one of the most important tasks of the fire investigator. Using a variety of media, the investigator records information regarding the incident that will be essential when it comes time to recall the conditions found and to provide documentation of the fire scene and the evidence collected.

Documentation of fires and explosions is essential because most scenes will be altered or destroyed after the investigation. Investigators should remember that they may not get a second opportunity to collect information or evidence once their scene examination is completed. Good documentation is a very effective tool in any investigation.

##  Documenting the Incident

In fire and explosion investigations, the most common methods used to document the investigative findings are:

- Field notes
- Sketches
- Drawing, diagrams, and maps
- Photographs
- Videotapes
- Tape recordings

## Field Notes

Field notes are the most common type of documentation used by fire investigators. They provide a written record of the investigator's observations and findings during the investigation. These notes will be used to develop the final incident report and may be used to provide background information when the investigator is called to testify in a criminal or civil trial. In most cases, the field notes are handwritten by the investigator while he is at the fire scene or continuing the investigation beyond the scene. At the conclusion of the investigation and after the incident report is completed, some investigators destroy their notes while others save them in the case file. The investigator should follow the procedures of his jurisdiction regarding the handling of notes. If the notes are destroyed, the investigator is open to allegations that case information beneficial to the defendant was deliberately destroyed. Such allegations could have a negative impact on the outcome of the investigator's case. It is important to remember that if you do retain notes, they are subject to subpoena should the case be brought to civil or criminal trial.

Good field notes will be accurate, complete, concise, and in the format required by the jurisdiction or employer of the investigator.

> A system for the collection of investigative information is found in NFPA 906, *Guide for Fire Incident Field Notes*. While the investigator may not always require the level of detail found in these forms, they provide an excellent overview of the type of information that should be collected. Using a system, such as presented in NFPA 906, will also assist in the organization of other types of incident documentation such as photographs, sketches, evidence collected, and interview data. The NFPA 906 forms are provided in Appendix C to assist the investigator in the development of his own field note format. The investigator should not be restricted by the design of the forms.

Organization is critical for the accurate development of a report and compilation of a complete and organized case file. The information contained in the field notes can also be used to provide an overview of the incident for statistical purposes for fire department records. Whatever format is used, the fire investigator must develop enough information so that a report can be prepared and the investigative findings can be accurately reported and communicated.

While developing their notes, fire investigators should take the following into consideration:

- Include the date and only the information that is pertinent to the investigation.

- Avoid the inclusion of personal comments or opinions.

- Record only facts and actual observations related to the fire scene.

- Do not mix information from different incidents in field notes— compile a separate set of notes for each incident.

- Be complete— you may not get the chance to get the information again.

- Remember that field notes may be read by others, including a jury.

- Use a systematic and consistent method when writing field notes. This will enable the investigator to recall and interpret them at a later date.

## Sketches

According to NFPA 921, the simple sketch is the minimum drawing the fire investigator should develop as part of an investigation. A sketch of a fire scene is completed while the investigator is on the scene. The sketch is intended to provide a graphical representation of the scene that is proportional but not necessarily to scale. The fire investigator should use common symbols in the sketch and provide a legend explaining the symbols used. Whatever symbols the investigator chooses to use, they should be used consistently in all sketches and drawings related to the investigation. Table 9.1 provides common symbols used in fire protection drawings that may help clarify sketches and drawings.

The basic sketch completed by the fire investigator should provide an overall representation of the fire scene and may include the following:

- Layout of the fire scene

- Overall measurements of the building or scene and individual rooms within a building

- Location of furniture and other contents

- Locations of witnesses

- Points where photographs were taken by the fire investigator

- The location of important information or evidence related to the investigation

- Location and position of any fire victims

- Entry points into the fire scene

- Nearby paths or roads

- Burn patterns

- Area/point of origin

- Fire travel

- Smoke patterns and travel

- Fire detection or suppression systems and devices

- Suppression activities

- Windows or openings

- Location of evidence

## Table 9.1
## Typical Plan Sketch Symbols

Access Features, Assessment Features, Ventilation Features, and Utility Shutoffs

△ FD Fire Department Access Point

△ K Fire Department Key Box

△ RA Roof Access

△ AP Fire Alarm Annunciator Panel

△ RP Fire Alarm Reset Panel

△ CP Fire Alarm Voice Communication Panel

△ WB Sprinkler System Water Flow Bell

△ SV Smoke Vent

Utility Shutoffs

△ E Electric Shutoff

△ W Domestic Water Shutoff

△ G Gas Shutoff

△ LPG LP-Gas Shutoff

△ NG Natural Gas Shutoff

△ CNG Compressed Natural Gas Shutoff

△ SP Smoke Control and Pressurization Panel

△ SL Skylight

◇ Detection/Extinguishing Equipment

◇ HD Heat Detector

◇ FS Flow Switch (Water)

◇ TS Tamper Switch

◇ DC Dry Chemical System

◇ WC Wet Chemical System

◇ CA Clean Agent System

◇ DD Duct Detector

◇ SD Smoke Detector

◇ PS Manual Pull Station

◇ HL Halon System

◇ CO₂ $CO_2$ System

◇ FO Foam System

◇ BSD Beam Smoke Detector

○ PIV Post-Indicator Valve

○ ZV Sprinkler Zone Valve

○ WH Wall Hydrant

○ TC Inspector's Test Connection

○ FDC Fire Department Connection

○ WT Water Tank

○ RV Riser Valve

○ HC Hose Cabinet or Connection

○ TH Test Header (Fire Pump)

○ FH Fire Hydrant

○ DS Drafting Site

Equipment Rooms

▢ AC Air-Conditioning Equipment Room

▢ EG Emergency Generator Room

▢ EE Elevator Equipment Room

▢ FP Fire Pump Room

Fire scene sketches are not intended to be works of art. They should, however, accurately capture critical incident information in a graphical format that is easy to understand. A key consideration when developing sketches for use in an investigation is to keep them uncluttered. It is better to use multiple sketches to show various information than to try to include too much on one drawing and make it very hard to read and understand. Figure 9.1 shows several examples of fire scene sketches and describes the various orientations that can be used to present the information.

Grid paper (graph paper) is an excellent aid to the fire investigator while performing this task. The grid on the paper can be assigned a scale, and the drawing can be completed to an approximate scale very easily. The investigator may also use items, such as straight edges or measuring devices and clipboards, to assist in the development of the sketch. Figure 9.2 gives an example of a completed sketch where graph paper was used.

Whatever method the investigator decides to use to produce field sketches of the fire scene, he will have to also use a reliable means of locating items on the sketch. The collection of this information will be essential in the reconstruction of the scene, documentation of evidence collection points, and the development of more detailed diagrams after the scene examination is completed.

## Drawings, Diagrams, and Maps

For many investigations, the simple sketch completed on the fire scene may not be adequate. The fire investigator may find it necessary to develop a formal set of drawings, diagrams, or maps that present detailed information related to the fire. Drawings may depict the floor plan of a fire building and have various levels of detail regarding the building, its construction, and contents. Information regarding the fire can then be placed on the drawing and several versions generated showing different types of data.

Detailed diagrams of specific areas or systems involved in the fire may be needed to provide necessary details for the investigation. Street maps may be used to show the location of the incident relative to the surrounding area, including nearby streets, the locations of other facilities, water sup-

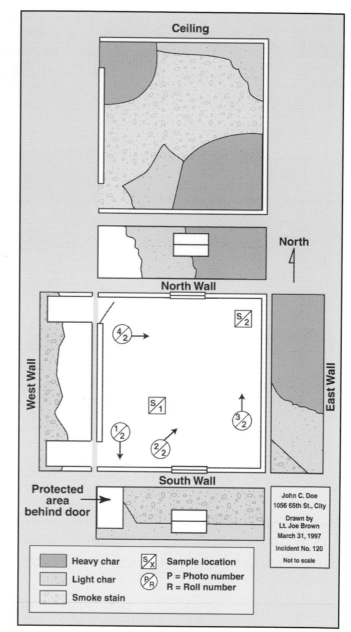

**Figure 9.1** Example of a fire scene sketch.

ply, etc. Maps can also be used to show the location of fire suppression units and where specific suppression activities took place.

For incidents that affect large areas, such as explosions or hazardous materials spills or releases, maps can be used to depict the damaged or affected area, weather conditions (e.g. wind direction), or any other specialized information. Investigative information may be overlaid on existing drawings, maps, and aerial photographs, or they can be developed from scratch (Figure 9.3).

Until the recent proliferation of personal computers, most of these items were hand drawn using

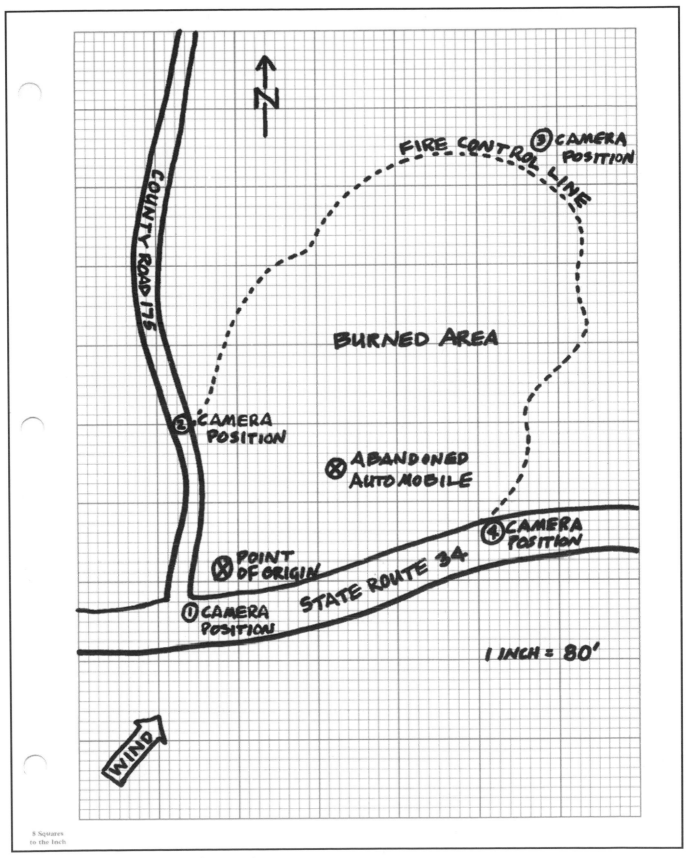

**Figure 9.2** Shows fire scene sketch on graph paper.

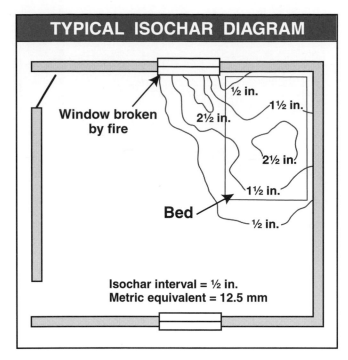

## TYPICAL ISOCHAR DIAGRAM

Window broken by fire

½ in.

1½ in.

2½ in.

2½ in.

1½ in.

Bed

½ in.

Isochar interval = ½ in.
Metric equivalent = 12.5 mm

**Figure 9.3** Typical Isochar diagram, which shows lines connecting points of equal char depth.

drafting equipment such as an engineer or architect would use. This equipment was specialized and difficult to use without training. Today, however, there are many computer-based drawing and CAD (computer-aided design) programs that allow the investigator to develop high-quality materials with very little special training (Figure 9.4).

In addition to developing their own materials, fire investigators may find drawings of a building used in the plan review, permit process, and construction helpful in documenting the incident. These drawings will normally provide a great deal of information and detail about a building and its systems. The fire investigator using building plans for the basis of a drawing should verify the accuracy of the plan and include any changes to the building made during construction or renovation. Table 9.2 gives an overview of the types of materials that might be available for larger buildings. Street maps

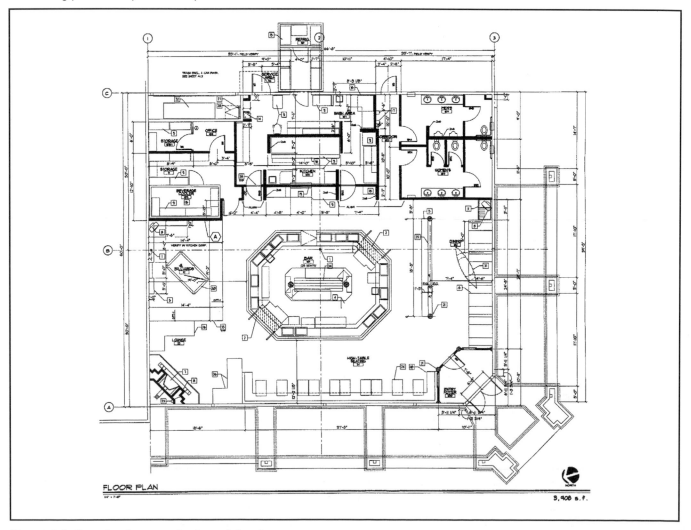

FLOOR PLAN

**Figure 9.4** Sample of a computer-based drawing.

| Type | Information | Discipline |
| --- | --- | --- |
| Topographical | Shows the various grade of the land | Surveyor |
| Site Plan | Shows the structure on the property with sewer, water, electrical distributions to the structure | Civil Engineer |
| Floor Plan | Shows the walls and rooms of structure as if you were looking down on it | Architect |
| Plumbing | Layout and size of piping for fresh and waste water | Mechanical Engineer |
| Electrical | Size and arrangement of service entrance, switches and outlets, fixed electrical appliances | Electrical Engineer |
| Mechanical | HVAC system | Mechanical Engineer |
| Sprinkler/Fire Alarm | Self-explanatory | Fire Protection Engineer |
| Structural | Frame of building | Structural Engineer |
| Elevations | Shows interior/exterior walls | Architect |
| Cross Section | Shows what the inside of components look like if cut through | Architect |
| Details | Shows close-ups of complex areas | All Disciplines |

of the area involved in the incident may be available from the street or engineering department for the jurisdiction. These maps can be copied and incident information placed on them for use in the investigation.

Using high-quality equipment or computers to generate drawings, diagrams, or maps for the investigative report will only be successful if the fire investigator collects details and accurate information during the scene examination. Almost all the information included in specialized artwork related to the incident will be based on the basic sketches and field notes produced by the investigator while on the fire scene.

## Photographs

While field notes and sketches of the fire scene are important components of the documentation pro-

cess, photographs are the key components. A good photodocumentation of the fire scene will assist the fire investigator throughout the investigation. Photographs allow the investigator to show the scene as it was found and to follow the investigative process at the scene through the reconstruction of the area of origin and the collection of evidence. Because the fire scene is very fragile and easily disturbed, photographs allow the investigator to show the scene at various stages in the process and to preserve the information visible during these stages. When the investigation is completed and the scene is cleaned, reconstructed, or demolished, the photographs taken during the scene examination may be all that remains.

In very large organizations, the job of photographing the scene may be the job of a department photographer. However, most fire investigators will be responsible for taking their own photos while on the scene. To accomplish this task and to accurately depict and support their findings, fire investigators must have a basic understanding of the equipment and techniques used in fire investigation photography. The following sections discuss the operation of the camera and associated equipment, lighting, and film. They also discuss using this equipment to get quality photographs in the difficult circumstances present at the fire scene.

## Camera

The workhorse of fire investigation photography is the 35 mm single lens reflex (SLR) camera. The SLR camera allows the photographer to view and focus the subject through the lens that will take the picture. The SLR camera can be focused and allows the user to adjust various settings to obtain a properly focused and exposed photograph of the subject. SLRs also allow the user to interchange lenses and use an external flash unit, depending on the photo they are attempting to take. Figure 9.5 is a typical SLR camera with the components labeled.

Modern SLR cameras will have built-in metering that will provide the investigator with the camera settings needed for a correctly exposed photo. The most recent models available provide automatic focusing and exposure control features that make the cameras very easy to use under normal conditions. The difficult conditions found on the fire scene, including very low light and the lack of contrast between the surfaces being photographed, may require the investigator to use this camera on manual mode and control the settings himself. He will have to depend on his knowledge of photography and know how to maximize the image that is placed on the film.

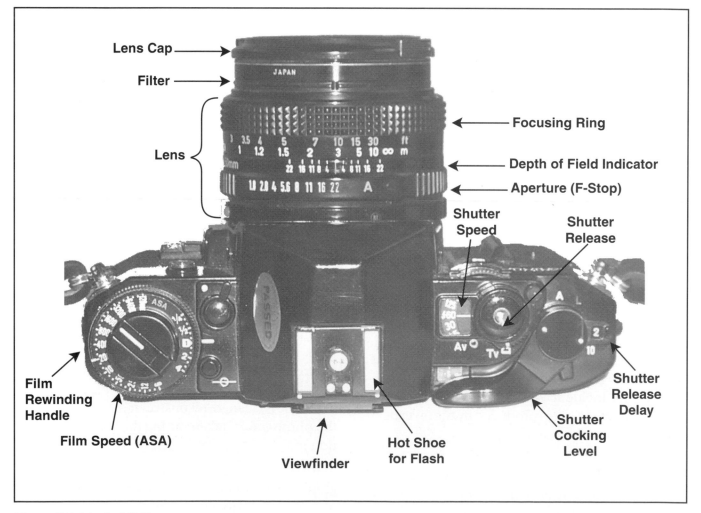

**Figure 9.5** A typical SLR camera.

Other types of cameras available include:

- Automatic focus (AF)
- 35-mm fixed-lens viewfinder cameras
- Instant cameras that provide a photo in a very short period of time
- Disposable cameras that are preloaded with film
- Digital cameras that use computer technology to capture and store an image

As the technology improves, different camera formats will prove to be useful tools for the fire investigator.

For an image to be made on the film in a camera, light must pass through the camera lens and be focused for a period of time on the film surface. For a properly exposed image to be made, the timing of the exposure and the size of the opening the light passes through must be controlled.

The camera shutter controls the amount of time the film is exposed to light. In SLR cameras, the shutter speed is adjustable from several seconds in length down to thousandths of a second. The aperture opening is the size of the opening in the camera lens that allows the light to pass through. The aperture opening on an SLR camera lens is adjustable and calibrated in adjustments known as f-numbers (sometimes called f-stop). The diameter of the aperture opening decreases as the f-number increases. The f-numbers will vary depending on the lens attached to the camera (Figure 9.6).

By manipulating the aperture opening and shutter speed, a properly exposed image can be made on the film's surface. The selection of the speed and f-number may be suggested by a metering system built into the camera and manually set by the photographer. This is automatically set by the camera in the newer electronically operated SLR cameras. Due to the complexity of the photographs he will be taking on the fire scene, the fire investigator should select a camera that will allow for user control of the adjustments. In difficult lighting situations, many investigators will "bracket" their shots. *Bracketing* involves taking a photograph at the setting recommended by the camera meter and then manually adjusting the exposure setting one or two f-numbers above and below the recommended exposure. Newer automatic cameras give the user the option of automatically bracketing each shot a specific range above and below the metered setting.

### Lenses

The function of the camera lens is to focus the light entering the camera onto the film. The SLR camera allows the use of many different lenses of various focal lengths depending on the photograph the user is attempting to take. The *focal length* of a lens is the distance behind the lens where light from an object is sharply focused when the lens is set to infinity. Variations in the focal length of a lens will determine the size of the image that is focused on the film. For SLR cameras, the 50 mm lens is considered the normal focal length because it provides an image on the film that is close to what the human eye would see. There are three ranges of lenses typically used with the SLR camera. They are:

- Normal— normal focal length— 50 to 55 mm (Figure 9.7)

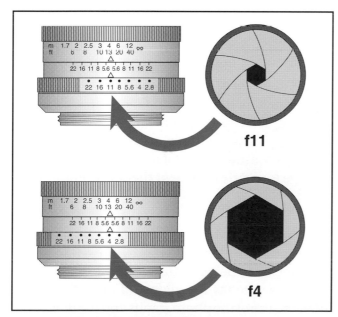

**Figure 9.6** Examples of different f-stops.

**Figure 9.7** Normal focal length.

- Wide angle— short focal length— 24 to 35 mm (Figure 9.8)
- Telephoto—long focal length— 105 mm and up (Figure 9.9)

If the investigator is going to take close-up shots of small objects with sharp detail, a macro lens or a close-up adapter for the camera lens is necessary. There are other special lenses such as the fish-eye with a very short focal length and a wide field of view (17 mm and 180°) and the super telephoto with a very long focal length and very narrow field of view (1 200 mm and 2°). These lenses are very expensive and have limited application in normal fire investigative photography.

The most versatile lens the fire investigator can use is the adjustable focal length or zoom lens with macro (close-up) capability. This lens allows the user to adjust the focal length of the lens to that desired for the photo being taken. Zoom lenses are available in a variety of ranges but the fire investigator is best served by one that is adjustable from wide angle to normal or just above normal focal lengths, such as 28 mm to 105 mm.

**Figure 9.8** Short focal length.

**Figure 9.9** Long focal length.

There are a variety of filters available for camera lenses. These filters provide a variety of effects in the resulting photograph. These effects can dramatically alter the photograph and thus reduce its effectiveness in representing the fire scene. For the fire investigator who is not an expert in photography, it is recommended that special filters be avoided. However, the use of a clear or UV filter on a camera lens to protect the lens from damage is acceptable and good practice.

### Depth of Field

Depth of field is a concept that fire investigators should understand and keep in mind as they photographically document the fire scene. The *depth of field* is the range that is in focus in front of and behind the subject of the photograph (Figure 9.10).

The factors that affect the depth of field of a photograph include:

- Aperture setting—the depth of field increases as the aperture opening decreases (f-number increases).
- Focal length of lens—shorter focal length lenses have longer depths of field. Wide-angle lenses have deeper depth of field than do telephoto lenses.
- Distance to the subject—the longer the distance between the camera and the subject, the greater the depth of field.
- The depth of field will be greater behind the subject than in front of it. By setting the point of focus approximately one-third the distance into the overall image, the investigator will place the objects of interest into the area that is in focus.

The detail of a fire scene photograph will be determined by the depth of field. The greater the depth of field, the more detail the viewer of the photograph will see. This is another concept fire investigators must understand and use to their advantage to clearly represent the fire scene.

### Film

The film used in a 35-mm camera is the medium that captures the image seen by the camera lens and allows it to be reproduced. The common types of film used today are slide and print films. The intended use of the photographs and the policies of the organization should determine the type of film

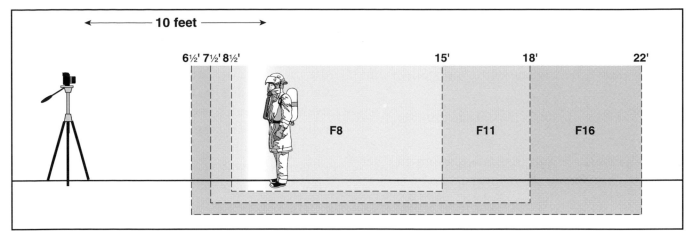

**Figure 9.10** Example of depth of field.

used by the fire investigator. Most investigative work should taken in color print film. Color print film records a wider range of contrasts and is less sensitive to variations in exposure settings; processing labs do automatic color correction. It is more forgiving than slide film and the resulting prints are fairly easy to view and use.

Color slides must be viewed using a projector or viewer. They are best used where the shots being taken may need to be viewed by many people. Slide films will provide sharper contrast in very low contrast situations, but as already stated, these films are very sensitive to variations in exposure settings.

The rated speed of film indicates its sensitivity to light. Film speed is set in accordance with standards established by the International Standards Organization (ISO). The higher the ISO rating of the film, the more sensitive it will be to light. The more sensitive the film is, the less light required for proper exposure. Film is available in the following speed ranges:

- Slow speed — films with speeds ranging from 25 to 100. These films are very good for use outside with natural lighting. These films will provide a sharper image as the size of the print is increased. They are not generally used for interior fire scene work.

- Medium speed—films with speeds ranging from 125 to 200. Useful for general purpose photography where normal lighting is available. Film usually has a fine grain, so good quality enlargements can be made. As with slow film, the speed may not be suitable for interior fire investigation photography.

- High speed — films with speeds ranging from 400 to 1,600. Provides the most versatility under varying lighting conditions. Recent advances in the technology provide sharper images and finer grain than with older high-speed films. This allows for sharp enlargements from the negatives. High-speed film is the choice of most fire investigators for interior photographs of a fire scene.

Modern cameras are provided with sensing devices that automatically set the proper film speed based on information coded on the film. Film with this information coded on it is designated with a DX on the container. With older camera bodies, the film speed must be set manually prior to making exposures.

The speed of the film used will determine the appropriate shutter speeds and aperture settings necessary for properly exposed photographs. Higher speeds will allow for the use of smaller aperture openings and thus provide a greater depth of field under low light conditions. As you will see later in this chapter, the film speed will also effect the use of electronic flash units where the available light is not adequate for taking a photograph.

Most fire investigators find that using the same film for the majority of their investigative work allows them to become familiar with the specific characteristics of the product and take consistent photographs from fire scene to fire scene. Fire investigators should experiment with the products that are available and select the one that provides them with the best overall results.

The proper storage and handling of film, unexposed and exposed, is an important issue for the fire investigator. Photographic film is perishable and can be damaged by a variety of external factors. The fire investigator should keep the following factors in mind when it comes to the storage and handling of film:

- Do not open the original package until ready to use the film.

- Use the film promptly (definitely before the "Develop Before" date on the package).

- Keep the temperature low. For best results, keep film temperature below 75°F (24°C).

- Avoid any excessive heat. Do not store film in the glove compartment or trunk of a car. Use insulated bag or cooler for storage in a vehicle during hot weather.

- Do not subject film to high relative humidity. Ideally, film should be stored in locations with a relative humidity below 50 percent.

- Protect film from exposure to X rays. Checked and carry-on bags on commercial airlines are all subject to X-ray examination. It is recommended that loaded cameras and film supplies be hand checked if you must carry them on an airplane. While a single exposure may not damage the film, repeated exposure may fog unprocessed film. Higher speed films are more susceptible to damage from X rays.

Once film is exposed, it should be processed promptly for best results. Negative film is printed on photographic paper, and slide film is mounted in slide holders for use. Once processed, the images can be enlarged and with special processing even digitized so that they can be used with a computer. Most investigators have a single copy of the images produced and select individual photographs for duplication for their report. The original negatives or slides should be stored in such a way that they will not be damaged or lost should they be needed later in the investigation.

## Lighting

The most natural lighting available for a photograph is from the sun. In fire investigation photography, however, natural light will rarely be adequate for the job. The investigator must almost always rely on an artificial source of light while documenting the scene. For this purpose, a good electronic flash unit that is compatible with the camera being used is essential. Other artificial sources of light at the investigative scene include floodlights or the lighting provided in a building. These sources, however, may not always be available and should not be relied on. The investigator may also carry a flashlight that can be used to focus the camera on objects in areas with very little light and allow them to view settings on the camera. If this technique is used, the focusing light should be turned off prior to taking the photo to avoid discoloration.

---

Guide numbers are based on the speed of the film and the BCPS output of the flash being used. Guide numbers are provided by film manufacturers and are normally available for the specific flash unit the investigator is using. Table 9.3 is an example of the data provided by a film manufacturer for one of its product lines.

To use this table, the fire investigator would select the BCPS output that was closest to the flash he was using and read the guide number for the speed of the film in the camera.

To calculate the recommended f-number setting for a photograph, the investigator would use the following equation:

$$f = G/x$$

Where:

f = f-number

G = guide number

x = distance from flash to subject

The following calculations are examples for an investigator using a flash unit with a rated output of 2,800 BCPS to photograph an object that is 20 feet (6.1 m) from the flash unit for several speeds of film:

For ISO 100 film, the guide number from the table would be 120 or 36 for SI.

f = 120/20 ft    or using SI   f = 36/6.1 m

f = 6                                        f = 5.9

For ISO 400 film, the guide number from the table would be 240 or 70 for SI.

f = 240/20 ft    or using SI   f = 70/6.1 m

f = 12                                       f = 11.6

For ISO 800 film, the guide number from the table would be 330 or 100 for SI.

f = 330/20 ft    or using SI   f = 100/6.1 m

f = 16.5                                     f = 16.4

## Table 9.3
## Guide Number Distances in Feet/Meters

| Unit Output (BCPS)* | Guide Number Distances in Feet/Meters | | | |
|---|---|---|---|---|
| | Gold 100 | Gold 200 | Gold 400 | Gold Max |
| 350 | 40/12 | 60/18 | 85/25 | 120/36 |
| 500 | 50/15 | 70/21 | 100/30 | 140/42 |
| 700 | 60/18 | 85/25 | 120/36 | 170/50 |
| 1000 | 70/21 | 100/30 | 140/42 | 200/60 |
| 1400 | 85/25 | 120/36 | 170/50 | 240/70 |
| 2000 | 100/30 | 140/42 | 200/60 | 280/85 |
| 2800 | 120/36 | 170/50 | 240/70 | 330/100 |
| 4000 | 140/42 | 200/60 | 280/85 | 400/120 |
| 5600 | 170/50 | 240/70 | 340/106 | 470/140 |
| 8000 | 200/60 | 280/85 | 400/120 | 560/170 |

*BCPS – beam candlepower seconds

Source: Eastman Kodak Company, Technical Data/Color Negative Film, Kodak Publication No. E-15, January 1997

The electronic flash selected for use by the fire investigator should have an output that is sufficient to light a darkened fire room and provide sufficient detail in the photograph. The low output units built into many cameras may not meet the needs of the fire investigator. The power output for electronic flash units is reported in beam candlepower seconds (BCPS) units. The higher the BCPS output, the more powerful the flash. If the BCPS output of a flash is known, the photographer can use the guide number for the film being used to determine the approximate f-number setting required to obtain a properly exposed photograph. The f-number is calculated by dividing the guide number by the flash-to-subject distance. (See the sidebar on page 118 for examples of these calculations and guide number data from a major film maker.) Many investigators have found the use of automatic focus camera systems very helpful in dark and inaccessible scenes.

So that the user will not have to constantly remember the flash output, guide numbers, and solve equations, flash units normally have a calculator dial built into them that allows the photographer to set the speed of the film being used and input the distance to the subject to obtain the suggested f-number. The fire investigator will have to experiment with the type of film and settings necessary to obtain good results on the fire scene. Many situations may require the use of manual settings on the flash to gain the maximum light output and require increasing the aperture opening by 1 or 2 f-numbers beyond the recommended setting. Thus, if the recommended aperture setting is f-5.6 for a shot that is approximately 30 feet (9 m) from the camera, the investigator should consider a setting of f-4 or f-2.8. In very difficult situations, such as heavily charred objects, the investigator should bracket the shots by taking one at the recommended setting and then additional shots at aperture settings both above and below the initial setting. When taking flash photos using large aperture openings (smaller f-numbers), the investigator should keep in mind that the depth of field of the photo will be limited. Care must be taken to ensure that all the information the investigator wants included in the photo is in focus.

Recent technology in the camera industry has yielded systems where the camera and flash unit are connected electronically and provide each other with data that will provide an optimal exposure for any given photograph. These units will have to be experimented with under the extreme conditions found at the fire scene to determine if they provide acceptable results. Typically, automatic units will have difficulties with settings where there are large bright or very dark areas. In the bright background, the automatic flash may provide settings that underexpose the photo. Large dark areas, such as burned out rooms, may end up overexposed. Most of these units also allow the user to manually operate the camera and input desired settings for shutter speed and aperture opening to compensate for extreme conditions.

When using a flash, the investigator may want to consider removing the flash from the camera body and holding it at an angle to the camera body. This will help to reduce light reflection, will increase the perception of depth, and provide a better representation of the textures of the surfaces being photographed. This is very useful when photographing heavy char. Another method used to reduce glare or reflection created when the flash is pointed directly at a subject is bounce flash. In this case, the flash is directed at the ceiling above the subject and then the photo is taken. The light bounces off the ceiling and provides a softer image of the subject. This technique will not work well in fire blackened areas.

A technique used to photograph a large area under low light conditions is painting with light (Figure 9.11). The photograph is produced by placing the camera on a tripod (or some other very stable position) and locking the shutter in the open position. The flash unit is then operated from several locations just outside of the field of view of the lens to illuminate the scene. Gently placing an opaque cover in front of the lens between flash operations will reduce extraneous light and movement on the image.

 ## Photographing the Fire Scene

Good photographs of the fire scene are an essential part of the scene documentation and investigation of any fire. Prior to taking any shots at the fire scene, the fire investigator should establish a photo log so that all shots taken can be identified and accounted for. Form 906-8 from NFPA 906, *Guide for Fire Incident Field Notes (1998),* provides a format for such a log. This form or one of similar design will be a major assistance to the investigator if the photos are ever used in court. The first shot on each roll of film used at a fire scene should identify the fire, investigator, and the date and time the roll was used. This practice will again help to identify the photographs should they ever be used in court proceedings.

To document the fire scene, the fire investigator should begin on the outside of the fire building. These shots should establish the location of the scene, any landmarks near the building, how the structure was located on the property, and the entrance of any utilities that might be visible. A structure should be photographed from all corners and if necessary, all sides to provide a record of the configuration of the building and related structural elements. During this period, the investigator may also take photographs of items found on the exterior of the building that might have a factor in the origin and cause of the fire. Apparent routes of fire travel should also be photographed.

When the exterior of the structure is completed, the investigator should move to the interior and

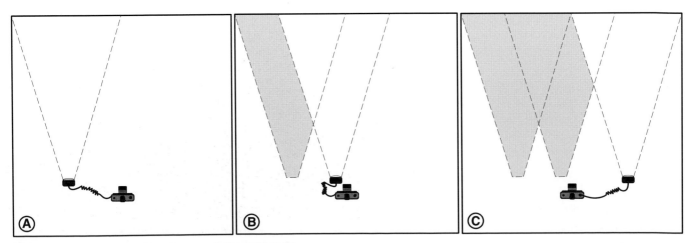

**Figure 9.11** This illustration shows painting with light.

begin photographing all the rooms in proximity to the fire as well as those rooms that were impacted by the fire (Figure 9.12). In a single-family home, this would probably be all the rooms in the structure. In a large high-rise office complex, it would be only the fire room and those adjacent spaces impacted by the fire or relevant to the investigation. For documentation purposes, the entrance to a room and all walls enclosing it should be photographed. Windows, doors, and contents should be visible in these photographs.

As the investigation progresses, additional photographs of the investigative process should be taken. As the area of origin is excavated and reconstructed, each step should be photographed. Any evidence collected should also be documented.

Other subjects for photographs at the fire scene include:

- Utility entrances to the building, including meters and controls

- Any appliances in or near the area of origin, including the position of controls

- Any victims and their locations

- Photographs that illustrate the witness's view of the scene

Fire investigators must remember that they may have only one chance to obtain photographs that document the fire scene. They should take as many photographs while on the scene as they need to fully document the fire and the scene in general.

## Video Cameras

While still photography is the primary means for documenting the fire scene, video cameras can play a role in the documentation. Where appropriate, video should be used to supplement the still photographs made during the investigation. The best applications of video are for documenting:

- Complex structures or floor plans

- Building contents

- Collection of special evidence where the sequence of collection may be important to document

- Interviews

The best application is to depict any event or sequence that takes place over a reasonable period of time where the sequencing of the events is im-

**Figure 9.12** Investigator photographs interior of structure.

portant to the investigation. The overuse of this medium should be avoided. Hours of tape without a specific purpose can actually hurt a case by boring the viewers.

When using video, avoid recording all extraneous comments. Caution should also be used so that the images recorded provide an accurate representation of the scene.

## Portable Tape Recorders

Voice tapes can also be useful tools for the investigator. Many investigators use tape recorders as a means for taking their field notes while in the fire building. These tapes are then transcribed into written notes. Tape recorders can also be useful during interviews as an adjunct to the investigator's notes. If the investigator plans to use a recorder in an interview, he must comply with the laws and policies of his jurisdiction and organization related to obtaining permission to record the interview.

As with any equipment, the fire investigator should not rely only on the recorder to capture critical investigative information. Always have a backup on paper or some other medium.

## ◆ Conclusion

The documentation of the fire or explosion scene is an essential step in the investigation of the incident. Good documentation allows investigators to illustrate their findings in their report and, if necessary, to present their findings in the courtroom months or years later. Using notes, photographs, drawings, and maps to document the scene is a skill every investigator must attain and use throughout his career.

# Chapter 10

# Evidence Collection and Preservation

## Performance Objectives

This chapter provides information that addresses performance objectives described in NFPA 1033, *Standard for Professional Qualifications for Fire Investigator*, particularly those referenced in the following sections:

**Chapter 3 Fire Investigator**
**3-4.2 a and b**
**3-4.4 a and b**

The job of a fire investigator is to collect and analyze information regarding a fire or explosion so that he can determine the origin and cause and the responsible party, if there is one. The information the investigator collects and analyzes is called *evidence*. Evidence can be in many forms such as the observations of witnesses, statements from building occupants or from the firefighters who fought the fire, and records and documents obtained after the fire. This chapter focuses on the identification, documentation, collection, preservation, and handling of physical evidence. NFPA 921, *Guide for Fire and Explosion Investigations*, defines *physical evidence* as "... any tangible item that tends to prove or disprove a particular fact or issue."

The term *evidence* is most commonly used in relation to criminal investigations where materials such as incendiary devices, ignitable liquid trailers or pools, and fingerprints might be critical to proving a case. Fire investigators must keep in mind that many accidental fires result in legal action of a civil nature. Thus, the results of their investigations into these noncriminal incidents may be used in court, and the materials they identify and collect as evidence could have a significant impact in these trials. Care must be taken when investigating these noncriminal cases and with the related materials that will be used as evidence in these cases.

One of the primary reasons that the fire investigator conducts a scene examination is to identify and collect information that will assist in the determination of the origin, cause, and responsibility of a fire. The responsibility for the identification, collection, and submission of evidence for analysis may rest with the fire investigator who conducts the initial scene examination. While additional materials or items may be collected by investigators who view the scene after the initial examination, the first investigator has the best opportunity to identify and preserve critical incident-related information and materials.

The physical evidence found at a fire scene can be in many forms (Figures 10.1 a-d). They include the following:

- Liquids thought to be accelerants
- Broken glass
- Portions of incendiary or explosive devices
- Liquid containers
- Appliances involved in the ignition of the fire

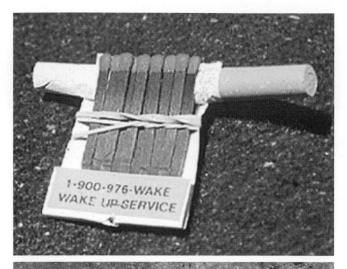

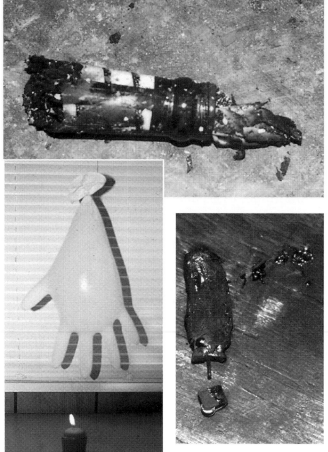

**Figures 10.1 a-d** Typical forms of physical evidence.

- Clothing/fabric
- Tire or foot impressions
- Tool marks on doors or windows
- Bodily fluids
- Cigarette butts
- Papers or documents

- Samples of charred wood, carpet, or other fuel involved in the fire
- Paint
- Hair/fiber
- Metal objects

The decision to collect materials at the fire scene is based on the relationship of those materials to the incident being investigated. When making the decision to collect physical evidence, the fire investigator must ask if the material can help to determine how the fire was ignited and how it spread once it started. The physical evidence, such as a device used to set the fire or materials that show specific fire patterns, may be directly related to the fire. The physical evidence collected may also be exclusionary in nature. *Exclusionary evidence* is collected to show that a particular device or scenario can be ruled out with relation to the ignition or fire spread scenario. Most evidence collected at the fire scene will be in the form of artifacts. *Artifacts* are the remains of materials involved in the fire that are in some way related to ignition, development, or spread of the fire or explosion.

Evidence collected by the fire investigator becomes the data used to develop and test opinions or hypotheses regarding the cause of the fire or explosion being investigated. This is related to the systematic methodology concept that was presented in Chapter 7, "Examining the Fire Scene." During the scene examination, the fire investigator will be called upon to:

- Recognize items of importance to the investigation.
- Protect those items.
- Document the location of items.
- State when items were discovered.
- Collect items to conduct analysis, and ensure that they are preserved for future use.

Previous chapters in this manual have discussed security and methods for protecting materials from additional damage during operations at the scene and documentation methods. This chapter discusses the recognition of potential evidence and methods that the fire investigator can use to properly collect, preserve, and secure these materials at the fire scene. The chapter also discusses the precautions necessary to prevent the materials collected from being contaminated or spoiled during or after their removal from the fire scene. The interpretation of the evidence collected, the subsequent analysis, and the use of the materials and related analysis in presentations will be discussed in later chapters.

 ## The Documentation and Collection of Physical Evidence

As the fire investigator conducts the scene examination, discussed in Chapters 12 and 13, he should constantly be looking for any physical evidence that will assist in the determination of the origin and cause of a fire or explosion. As previously discussed, the initial investigator on the scene has the best opportunity to recover these items in the most usable condition. As part of the investigative process, materials identified as potential evidence should be secured and protected from further damage until they can be documented and collected for preservation and laboratory analysis, if necessary. In general, it is recommended that samples be taken as soon as possible during the examination of the scene to prevent damage, destruction, or the complete loss of the materials.

Fire investigators should follow the procedures established by their agency regarding the collection of physical evidence. If materials are to be sent to a forensic laboratory for analysis, the sampling protocols of that laboratory should also be followed. Most laboratories will provide guidance on the type of packaging in which they would like to receive materials and the amount of the material they will need to conduct their analysis. ASTM E 1188-95, *Standard Practice for Collection and Preservation of Information and Physical Items by a Technical Investigator*, also provides useful information for the fire investigator.

Scene documentation should include notes, diagrams, and photographs that show the following:

- Specific location of the item at the scene

- Identity of the investigator who located the material and removed it from the scene

- Description of the evidence and any unique markings or labeling

- Date and time the item was located and/or collected

- Photographs showing the item as it was found and its relationship to other items

When the overall scene has been documented, each item of physical evidence that will be collected should be documented before it is disturbed.

Most investigators assign a unique identifier to each item of evidence located and collected. This identifier becomes important later if the custody of the material is transferred or the material is placed in storage. When an item of evidence is tagged and assigned an identification number, this number should be recorded in the investigation notes and on the scene diagram. The fire investigator should also generate an evidence log sheet that lists each item collected. (See NFPA 906, *Guide for Fire Incident Field Notes*, for an example of an evidence log sheet [Form 906-7]). The log will become an essential portion of the documentation of the chain of custody of evidence collected during the investigation.

### Chain of Custody

Maintaining the chain of custody of evidence is an essential function of the fire investigator. The chain of custody tracks an item of evidence from the time it is found until it is ultimately disposed of or returned (Figure 10.2). Each person who has possession of an item of evidence must be able to attest to the fact that the item was not subject to tampering or contamination while it was in his custody. If, as in the case of a laboratory, the item was altered while in the technician's possession, the techni-

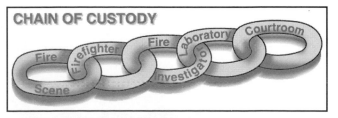

**Figure 10.2** The chain of custody must not be broken.

cian must be able to document what was done to it and provide results of the testing. Anytime the custody of an item changes, the following information should be obtained:

- Name and address of the current and prior custodian

- Description of any modification, handling, testing, or other alteration that occurred while in the custody of the current custodian

- Condition of the item or its packaging when it was transferred to the new custodian

While materials are being collected, packaged, and removed from the fire scene, the fire investigator should have a method to maintain the security of all items collected. Some investigators put evidence containers in a large toolbox provided with a lock; others may use a lockable box in their vehicle or the trunk of a car. Whatever method is used, investigators must be able to document that all materials collected were secure and in their custody until they were either securely stored or turned over to the custody of someone else (Figure 10.3).

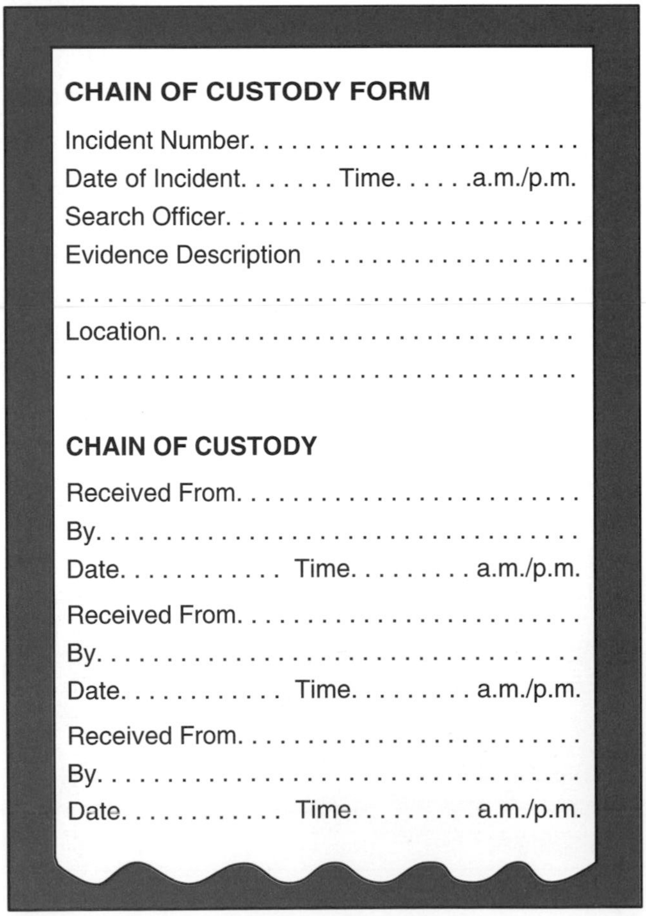

**CHAIN OF CUSTODY FORM**

Incident Number. . . . . . . . . . . . . . . . . . . . . .
Date of Incident. . . . . . . Time. . . . . .a.m./p.m.
Search Officer. . . . . . . . . . . . . . . . . . . . . . . .
Evidence Description . . . . . . . . . . . . . . . . . . .
. . . . . . . . . . . . . . . . . . . . . . . . . . . . . . . . . .
Location. . . . . . . . . . . . . . . . . . . . . . . . . . . . .
. . . . . . . . . . . . . . . . . . . . . . . . . . . . . . . . . .

**CHAIN OF CUSTODY**

Received From. . . . . . . . . . . . . . . . . . . . . . . .
By. . . . . . . . . . . . . . . . . . . . . . . . . . . . . . . . .
Date. . . . . . . . . . . Time. . . . . . . . a.m./p.m.
Received From. . . . . . . . . . . . . . . . . . . . . . . .
By. . . . . . . . . . . . . . . . . . . . . . . . . . . . . . . . .
Date. . . . . . . . . . . Time. . . . . . . . a.m./p.m.
Received From. . . . . . . . . . . . . . . . . . . . . . . .
By. . . . . . . . . . . . . . . . . . . . . . . . . . . . . . . . .
Date. . . . . . . . . . . Time. . . . . . . . a.m./p.m.

**Figure 10.3** A typical chain of custody form.

Using evidence tags or placing the information directly on the package in which the material is stored assists in tracking the item (Figure 10.4). The evidence log from NFPA 906 shows how to track the custody of items of evidence.

This chapter is primarily aimed at the collection and preservation of physical evidence. Chain of custody issues extend beyond the physical evidence collected at the fire scene. All incident-related evidence including photographs, documents of evidentiary value, and any other items that support the findings of the investigation could be covered by chain of custody. This, however, depends on the case and the policies of the jurisdiction the investigator represents. Once an item is considered to be evidence, it must be properly secured and the handling and transfer of custody documented from the time it comes into the investigator's possession until it is ultimately destroyed.

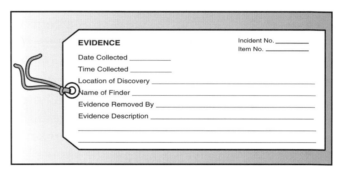

**Figure 10.4** Evidence tags identify the contents of the container.

## The Tools and Equipment Used to Collect Evidence

A list of tools used by the fire investigator was presented in Chapter 1, "Organization, Responsibilities, and Authority." Many of these tools are used in the collection of physical evidence at the fire scene. In addition to the hand tools previously listed, the fire investigator may also need larger tools such as hand shovels, garden-type rakes, hoes, crowbars, street brooms, and squeegees. These tools are used to clear away fire debris during the scene examination.

In addition to their tools, fire investigators should have a supply of containers and packaging materials for use in the collection and preservation of materials (Figure 10.5). In addition to containers, such as new paint cans, bags, and boxes, the investigator will need materials to assist in packaging

**Figure 10.5** One type of evidence container.

evidence when it is collected. These materials could include packaging tape, permanent markers to mark containers, filler material to pack glass containers and other fragile items to prevent breakage, and plain wrapping paper to protect materials such as glass sheets.

## Preventing Contamination

A major concern with regard to tools and equipment used in the collection of physical evidence is the prevention of cross contamination of the evidence being collected. To prevent this occurrence, the fire investigator must either use disposable items during the collection process or thoroughly clean all tools between each use. Contamination of the scene or samples being collected at the scene can occur from many sources. Some of the most common sources include:

• Tools used by firefighters and investigators at the scene

• Use of contaminated containers to store evidence

• Protective equipment worn by firefighters and investigators working at the scene

• Fuel-powered equipment used at the scene during suppression and investigation operations

New tools with rust-preventative coatings should be cleaned before use. During the investigation,

the tools used in the evidence collection process should be cleaned after each use with isopropyl alcohol or dish-washing detergent, flushed with clean water, and thoroughly dried using clean paper towels. This cleaning process must be conducted away from the scene so that it, in fact, does not contribute to any contamination. After the investigation, the investigator should use clean water to wash his tools and thoroughly dry the tools with paper towels to prevent rusting.

The protective equipment worn by firefighters and investigators is also a potential source of contamination at the fire scene. During the evidence collection process, investigators should be careful to clean their boots before entering an area where samples will be taken. As with tools, this can be accomplished by flushing with water and making sure that any foreign materials that may have been on their boots are removed. Another potential source of contamination are gloves used in fire fighting. The fire investigator should avoid using fire fighting gloves while collecting samples. During the actual collection process, a pair of unused latex exam gloves should be used. These gloves should be discarded after each use. Exam gloves, however, should NOT be placed in the container with the materials that were collected as evidence. To document the fact that new gloves were, in fact, used for each sample taken, some investigators number their gloves and the evidence containers sequentially as they are used. Then, they photograph the gloves with the sealed evidence container at the location the material was collected (Figure 10.6). The used gloves can then be discarded.

**Figure 10.6** Photographing the exam gloves with the evidence container is a good practice.

The use of fuel-powered tools on the fire scene is a potential source of contamination. After suppression operations are completed, the investigator should document where this type of equipment was used and whether it was refueled while it was in place. If the use of this type of equipment is necessary during the investigation, the investigator should take precautions to prevent contamination beyond the area in which the equipment is located. Investigators should wear different gloves if they have to fuel the device. They should take care with the fuel supply to avoid spillage on the fire debris or their clothing.

The key to minimizing potential evidence contamination or destruction is good scene security, the maintenance of the chain of custody of items collected, and the careful monitoring of those who enter the scene for possible contamination on their clothing, tools, or equipment. Careful cleaning of tools used during the collection process, or the use of disposable tools, will also reduce the potential for contamination.

## The Collection of Physical Evidence

How the materials are collected depends on what they are, their location, and the condition or state in which they are found. As the fire investigator begins to collect physical evidence, he will have to use his knowledge and skills to properly gather and preserve the evidence. If the item is intact and not covered with debris, the process is fairly simple — the item is documented, tagged, and placed into a container for transport from the scene. If the materials are covered with debris and possibly damaged as a result of the fire, the collection process is accomplished much like an archeologist would collect an ancient artifact. The item should be carefully exposed and documented and then carefully collected and packaged. When exposing the item, take care not to damage or disturb it.

The methods used for collection will depend on several factors that include:

- Physical state of the material — is it a solid, liquid, or gas?
- Potential for the material to evaporate or decompose after collection.

- How fragile the item being collected is, particularly if it is fire damaged.
- Size and physical characteristics of the item — is it very large or is it very small?

The objective of the fire investigator during this phase of the investigation is to collect and preserve material that will be used to support his opinions regarding the origin, cause, and responsibility of the incident. The collection of physical evidence should be accomplished following the policies and procedures of the fire investigator's organization or jurisdiction and the guidelines of the forensic laboratory that will perform any required analysis. For items or materials that are very fragile, very large, or in some other way difficult to collect, the fire investigator should consult with the forensic laboratory or a trained evidence technician for assistance.

## Evidence Collection Methods

Investigators must develop techniques for collecting and preserving physical evidence that they find at the scene of a fire or explosion. The primary objective of these methods is to prevent the material being collected from damage or deterioration. The investigator must always remember that the ultimate objective is to preserve the material so that a laboratory can examine and analyze it and that the material (evidence) can be viewed in any court proceedings that might result from the investigation. Thus, once the material is collected, it must be protected from contamination or destruction before it is analyzed or presented in court. Investigators will often have to improvise or use their best judgment when collecting and preserving evidence because of the variety of materials and circumstances they face in the course of conducting their investigations (Figures 10.7 a and b).

The following information provides fire investigators with basic information regarding the collection and preservation of typical materials they will encounter during a fire or explosion investigation.

## Body Fluids (Serological Evidence)

Body fluids found at the scene of a fire or explosion can be important items of evidence. They can be

**Figures 10.7 a and b** Different forms of evidence require different methods of collection.

used to show that an individual was at a scene or help to eliminate someone from consideration as a suspect in a crime. Examples of where serological evidence could be found at a fire scene include:

- Blood on a window broken to gain entry during an arson attempt

- Saliva on a threat letter

- Cigarette butt found at a fire scene

When dealing with any bodily fluids, the investigator should practice good infection control measures, including the use of personal protective equipment such as latex gloves, goggles or safety glasses to protect from splatters, hand washing

after handling body fluids, and thorough cleaning of any fluids that may get on clothing or footwear. **Always avoid direct contact with any body fluids.**

### Collection and Preservation

The investigator should carefully document existing patterns of blood spatters before any attempt to collect samples for laboratory analysis. This documentation should include close-up photographs of individual blood drops depicting the size, shape, and direction of movement.

Materials stained with blood or other body fluids that are to be collected should be air dried on clean paper in a dust-free location before packaging. Failure to dry samples could result in their destruction due to putrefaction. When dry, the material should be placed in an air-permeable container, such as a paper bag, and securely closed and sealed. Carefully wrap fragile samples, such as glass fragments with blood on them, in paper to prevent breakage and then securely package them in a bag or box. If a stain is found on an item or surface that is too large to submit intact, cut away or otherwise dismantle the stained portion. Package the sample using a box or wrapping it with heavy paper, and tape the package to prevent additional damage. Large pools of blood can be sampled by soaking up a sample with a gauze pad, air drying the pad, and placing the pad into an envelope for protection during transit to the lab. As a last resort, stains on very large surfaces can be scraped off the surface onto a sheet of clean white paper. The paper is then folded to enclose the particles and placed into a paper envelope for additional protection. Scrapings should not be placed directly into an envelope. Where blood stains are found on the ground, the stain and the material below it should be lifted intact and placed in an air-permeable container such as a pill box or cardboard box just large enough for the sample to fit into.

### Comparison Samples

Where possible, provide a comparison sample to the laboratory that will perform the analysis. Qualified medical personnel should collect comparison samples from live individuals or victims found at a scene. These samples should be preserved using the policies of the laboratory that will conduct the analysis.

## Hair

Hair found at a fire or explosion scene or on devices or items left behind by suspects can be used to assist in the investigation. One method used to obtain material for collection is sweeping or vacuuming. Hair found at the scene on a device is helpful because it is easily analyzed and resists biological decomposition, and it can provide investigative leads and associations. It is useful because it originated directly from the individual. Laboratory analysis can provide information regarding the following:

- Species the sample is from (for example, human, cat, or dog)
- Body area human hair is from (somatic area — scalp, facial, pubic, etc.)
- Racial origin of human hair
- DNA matches between samples from the scene and those of a suspect

### Collection and Preservation

To prevent damage, handle very carefully hair samples found at the scene of a fire or explosion. Clean tweezers can be used to remove the samples from where they are found. Be careful not to crush or break the individual hairs as they are being handled. Place recovered hair on a sheet of clean white paper. Then fold the paper around it. The folded paper containing the sample can then be sealed into an envelope and marked as required by the procedures of the investigator's jurisdiction. Clean sealable plastic or glass vials may also be used, with the hair being carefully placed in the container. As previously discussed, use caution not to crush or break any of the sample in the process. The vials are then sealed and properly marked or identified.

### Comparison Samples

Provide a comparison sample to the laboratory that will perform the analysis. There are very specific guidelines for collecting usable comparison samples of hair from an individual; therefore, the investigator should seek the guidance of a forensic laboratory for specific procedures.

## Fibers and Textiles

These items include articles of clothing or portions of the fabric used to construct a piece of clothing, carpet, and rope or string. These materials may range in size from large items such as articles of clothing or carpet samples to very small items such as fibers removed from a broken window or from an incendiary device. The composition and structure of samples found at a scene and on or with a suspect can be analyzed for similarities.

### Collection and Preservation

For larger samples, make sure that they are dry. Place the entire item in a box, bag, or packing tube. Seal and mark the container for identification. Smaller items must also be dry before packaging. When the material is dry, carefully place it in a glass or plastic vial using forceps if necessary. The method previously described for hair samples — wrapping the fibers in clean white paper and sealing into an envelope — could also be used here for smaller samples.

### Comparison Samples

Fibers and textiles are often collected to assist in the determination of the flame spread rating, flammability, and toxicity of a product. In these circumstances, multiple samples may be necessary. These samples will be collected and preserved as previously described.

## Tape

Individuals often use tape to construct or secure incendiary or explosive devices. As a result, tape may be found in the course of the scene examination. Laboratory examination of tape can provide information such as the type and manufacturer of the product. The laboratory can match torn or cut ends of tape to partial rolls recovered from a suspect. The laboratory can also provide information regarding the color, construction, and chemical composition for comparison with samples from other scenes or samples recovered from a suspect. The analysis and examination may also provide additional trace evidence such as fingerprints, hair, or fibers between the tape layers collected.

### Collection and Preservation

When found in layers as part of a device, cut the tape and lay it open rather than unwind it from the device. Place the material on a polyester sheet, such as a document protector, and package it in a bag or box to prevent additional damage. Single layers used to hold a device should be handled very carefully so as not to destroy any trace materials attached to the adhesive side. Package in the same fashion as the layered material.

### Comparison Samples

Rolls of tape suspected of being the source of the tape found on a device should be packaged to protect them from damage or contamination. These rolls are then provided to the laboratory with the sample removed from the scene.

## Glass

Glass from the scene of a fire or explosion may be collected to determine its composition and application; for example, was it from a bottle that was used as a Molotov cocktail or from a pane of window glass. Laboratory analysis can also determine whether fracture patterns were caused by heat, impact, high-velocity projectiles, or if a glass cutter was used. Analysis can also determine the direction of the force that resulted in a fracture.

### Collection and Preservation

The fire investigator assigned to collect glass as physical evidence must pay careful attention to the collection and packaging methods used at the incident scene. Before any broken glass is collected as evidence, the fire investigator should carefully document (see Chapter 9 for information regarding scene documentation) the location of all fragments that are to be collected. Marking glass fragments can be accomplished using tape or adhesive-backed note paper. Numbering each fragment collected will help with documentation. If the intent of collecting glass evidence is to determine the direction of force that resulted in the breakage of window glass, the following steps should be followed:

Step 1: Photograph all glass remaining in the broken window.

Step 2: Mark the remaining pieces as to their location in the frame — top, bottom, left side, right side, and the inside and outside surface.

Step 3: Remove and individually wrap each of these pieces before carefully packaging them to prevent further breakage.

Step 4: Identify and collect all available glass fragments from the broken window.

Step 5: Mark each fragment individually to indicate its location — inside or outside the building.

Step 6: Wrap and package each fragment.

Packaging techniques used for glass evidence will depend on the size of the fragment. The fire investigator must, however, use care to protect edges and also avoid scratching the surface. Tweezers or small pliers should be used to handle the fragments. Any thin edges that require preservation can be protected by embedding them in modeling clay or putty before the fragment is wrapped.

Mark large sections of glass as necessary, and wrap it in cotton or use padded packing material for protection. Place it in a sturdy box that can be sealed and marked as to its contents. Where control samples are to be submitted, they should be packaged separately.

Carefully handle small glass fragments to avoid any additional breakage. Place the fragments in containers such as pill boxes or small vials (glass or plastic). Pack these containers with cotton to prevent movement that could cause chipping during transit. The investigator should properly mark each container. **Do not use paper envelopes to package glass fragments.**

During the collection of glass evidence, the fire investigator should be alert to the presence of additional types of physical evidence that might be related to or attached to the glass being collected. For example, if the evidence is a broken window, the material used to hold the glass in place — putty or glazing compound or portions of the actual frame — might assist in the analysis. There could be tool marks on the window or wood splinters associated with the breakage. The investigator could also find fibers, hair, or blood on the glass. Any additional items discovered should be preserved using the proper technique for that material. Care should be taken to prevent this additional material from being destroyed or contaminated during the collection and packaging of the glass.

## Comparison Samples

If the glass collected is sent to a laboratory for impact analysis, the investigator should attempt to collect a comparison sample from as close as possible to the point of impact. If the material analyzed is from a bottle that contained flammable liquids and similar containers are found in the area, an unbroken bottle can be collected for comparison with the sample collected.

Glass comparison samples collected by the investigator should be packaged in the same way as described for glass evidence. Package comparison samples separately from the evidence. Carefully identify and mark comparison samples to prevent confusion at the laboratory.

# Wood

The collection of evidence composed of wood is a common task the fire investigator will be called on to accomplish. Items made of wood are frequently involved in fires and explosions. Wood evidence might be in the form of structural components such as paneling from the interior of a room; trim materials from around doors, windows, or floors; and window or door frames. On the other side of the size spectrum, the evidence could be very small in the form of wooden matches, splinters, or sawdust. Wood materials may be collected to determine the type of material of which the item is composed or for the analysis of burn patterns or tool marks found on the surface.

## Collection and Preservation

As with most materials, the collection and preservation of wood evidence depends on the physical size of the item and the intended use of the evidence. Entire sheets of wall paneling might be removed from the area of origin to show a particular burn pattern, or a section of a door frame may be collected to preserve a tool mark that is embedded in it. Carefully remove these large items from their location. Wrap them, if possible, to prevent additional damage, and tag, log, and transport them to a secure location. Large items may require the use of a truck for transport. Wrap smaller items in paper or plastic and then seal, tag, and secure until used.

Wet wood items are not intended for use as accelerant evidence. These should be air dried before

packaging to prevent the formation of mildew that could potentially destroy the evidentiary value of the item.

Place charred materials on a sheet of cardboard or similar material to prevent additional deterioration during the collection and packaging process. Dry the item, if necessary, and then wrap or pack it using packaging material or cotton batting to prevent motion during transport.

Small items, such as match sticks or splinters, may be placed in containers such as vials or pillboxes padded with cotton or other material to prevent damage in transit. These should be sealed and marked by the investigator.

# Paper

Paper-based materials the fire investigator may find could include:

- Materials used to wrap or hold a device
- Fireworks or dynamite tubing
- Paper products used as trailers to spread an intentional fire
- Paper matches
- Wet and/or burned documents found at the scene

Laboratory analysis can provide information as to the type and use of the sample, and where comparison samples are available indicate if there is a physical match between samples. Fingerprint evidence can also be obtained from paper products.

## Collection and Preservation

Where the paper product is suspected of being used as a trailer and may have ignitable liquids on it, handle the product as described in the section on accelerant evidence. Wet paper products that were not collected as accelerant evidence should be air dried before packaging to prevent the formation of mildew.

As always, it is recommended that the items be well documented before any effort to remove or collect them. For documents or materials with writing or printing on them, take close-up photographs. Place dry items in a container large enough to hold the material without damaging it. Place paper fragments in vials or small boxes.

Collect charred documents and other larger paper items by sliding them onto a cardboard sheet or flat stiff surface, such as a glass or metal plate. Collect remaining fragments of the document and package with the larger item. Use tweezers to handle the smaller fragments. The document and the material used to support it should then be placed into a cardboard box or other ridged container. Use cotton batting to protect the material during transport.

Paper materials found in a container, such as a wastebasket, drawer, or safety deposit box, should be left intact in the container. The entire container should be collected, packaged, and identified as evidence by the investigator.

Use tweezers to handle paper sheets that are dry, intact, and undamaged. These items should be collected as found. Do not fold, staple, or otherwise disturb the document. The investigator should initial and date the item in a noncritical location. Place the document into a protective covering such as a large envelope or plastic document protector. Then mark the outside of the container as evidence.

## Cigarette Evidence

Cigarettes or cigarette butts may be found at the incident scene. They may have value as evidence if they were used as part of an igniter for an incendiary device, were left by an intruder, or were involved as the source of the heat of ignition in an accidental fire. Laboratory analysis of cigarettes and butts can provide information as to the brand, type, flavor, length, and packaging. Analysis will not, however, be able to determine if cigarette evidence originated from a specific pack of the brand being analyzed. Laboratory analysis can also be conducted for the saliva and/or fingerprints that are on the butt.

### Collection and Preservation

As with other similar materials, handle cigarettes and butts with care so that any evidence that can be obtained is not destroyed. Document the items before they are disturbed. Investigators should not touch the items with their hands. They should use tweezers or a sheet of clean white paper to pick up the individual item from where it is found. Items that are wet should be air dried before packaging.

Package dry cigarettes and butts individually and place them in air-permeable envelopes. **The use of plastic bags is not recommended.** The investigator should identify and mark as evidence each envelope containing cigarettes.

## Components of an Electrical System

Portions of a building's, vehicle's, or equipment's electrical system may be collected to document its relationship to the ignition of a fire. In other words, the material may be collected to prove or disprove that the electrical system was involved in the ignition sequence. The fire investigator should make certain that the system is de-energized and that all sources of power to the system are disconnected before there is any effort to analyze or collect portions of an electrical system.

### Collection and Preservation

Components of an electrical system that have been exposed to heat and fire may be very fragile. The fire investigator should use caution while collecting these items to prevent damaging the items any more than they were when discovered. If possible, the experts who perform the analysis of the components should view, document, and possibly test (nondestructively) the system before it is dismantled for collection. If this is not possible, the items should be very carefully handled to prevent spoliation.

Before collection, all items of the system that will be collected should be thoroughly documented while still in place and undisturbed. The fire investigator should always collect enough of the material, such as wiring, so that it can be properly analyzed by the laboratory. If the item is an extension cord, the entire cord, the receptacle, and the junction box the receptacle is in should be collected intact. Where branch circuit wiring is to be collected, remove enough to show its condition. If practical, collect enough of the circuit so that wiring that is undamaged is also included.

When marking electrical evidence, the fire investigator should tag both ends of any wire collected—indicate the end connected to the device and the end connected to the electrical supply. In addition, the following may also need to be documented:

- The breaker to which the wire was connected

- The condition of the breaker as found
- The path of the wire between the breaker and the device to which it was connected

When it becomes necessary to collect a component of the electrical distribution system from a structure or vehicle, the component and attachment hardware should be removed intact and in the condition they were found, if possible. For example, investigators may opt to remove part of the structural elements (for example, wall studs) where components or devices are attached. Again, documentation before any collection effort is critical. The fire investigator should not operate controls, circuit breakers, or other components of an electrical system during the scene examination or while being collected and preserved.

Package any electrical system components collected as evidence in such a way that they will not be damaged during transport and storage. Place wiring in plastic bags, Wrap with paper or plastic, and seal with tape. Fire investigators should remember that heated wire becomes brittle at termination points or other points of interest. For this reason, do not bend the wire at these locations during packaging. Place panel boxes and other larger components in rigid containers, wrap with paper or plastic, and seal with tape. Package smaller components in boxes or bags, and pad them with cotton or paper to prevent movement in transit.

## Appliances and Equipment Involved in Ignition

When appliances or equipment are suspected of being involved in the ignition of the fire, they should be preserved as evidence. The fire investigator may request that the device be examined by specialized experts before it is moved, or investigators may seek a specialist's advice before collection. Even in fires that are accidental in origin, litigation may result involving such equipment, so the process of documentation, analysis, collection, and preservation is very important. In some cases, it may be appropriate that other parties be contacted before moving these devices. (See discussion on evidence spoliation in Chapter 8, "Securing the Fire Scene.") Where the conditions do not permit on-site analysis of a device, the fire investigator should proceed with collection and preservation efforts.

### Collection and Preservation

When practical and proper, the fire investigator should collect the entire device including the source of power, that is, electrical wiring or fuel lines connected to the device. For example, if the device being collected is a countertop kitchen appliance, the fire investigator would collect and preserve the appliance, the electrical cord attached to it, and possibly the electrical junction box or receptacle to which the cord was connected.

Large devices, such as a furnace from a building, should be protected in place for analysis. If controls or other components must be removed for preservation, investigators should collect enough of the component and related material, such as wiring or piping, to support their hypothesis regarding the device. The evidence should be maintained in the condition in which it was originally found. Switches and other types of controls should not be operated, and the equipment should be packaged to prevent damage during transport and storage.

## Ignitable Liquid Evidence

Ignitable liquid evidence is one of the most common types of evidence the fire investigator will collect at the fire scene. This evidence is collected and analyzed to determine whether an ignitable liquid is present in the sample. Ignitable liquids include flammable or combustible liquids. Ignitable liquids may be found at a fire scene because they either were used for normal functions conducted in the building or were used to intentionally accelerate a fire.

A *flammable liquid* is one with a flash point below 100°F (37.8°C). A *combustible liquid* is one with a flash point above 100°F (37.8°C). By their nature, flammable liquids vaporize easily at temperatures well below those found in a fire. However, unburned traces of these liquids can often be located during the scene examination. Because the samples are by nature volatile (easily vaporized), the collection and preservation methods used by the fire investigator will have a significant impact on the ability of the laboratory to perform the necessary tests for the materials.

### Collection and Preservation

The fire investigator must understand the physical properties of ignitable liquids to properly locate, collect, and preserve them as evidence. Properties that impact collection include:

- Liquids flow to low points on the surface on which they are poured and collect in pools at the low points and in any cracks or openings in the surface.

- Hydrocarbon-based ignitable liquids are less dense than water and will not mix with it (*immiscible*); therefore, they tend to float on the surface.

- Materials such as alcohol and acetone mix readily with water.

- Ignitable liquids are often absorbed by the materials on which they are poured such as floors, walls, or furnishings,

- When trapped in a porous surface, the liquid often remains intact after the fire and extinguishment efforts.

Due to the volatile nature of the material, collection efforts should be started as soon as possible after the fire. Specific indicators for sampling locations and the uses of adjuncts in the determination of sampling locations for ignitable liquids are discussed in Chapter 12. Ignitable liquid samples may be obtained in numerous forms including fire debris (such as carpet, cloth, wood, and flooring), pools of the suspect liquid on the surface of water used for extinguishment, and from containers found at or near the fire scene.

The fire investigator should package all materials collected as ignitable liquid evidence in airtight containers to prevent evaporation. The most common container used for this type of sample is a new paint can with airtight friction lids. Glass mason-type jars with tight-fitting lids are also acceptable but are fragile and must be handled with caution. Store all containers to prevent contamination before their use. Plastic bags, coffee cans with plastic lids, and other containers that leak vapors even when sealed should not be used for ignitable liquid samples. Containers used by the fire investigator should be approved by the forensic laboratory the sample will be sent to for analysis.

When collecting solid debris as accelerant evidence, the container should be filled to no more than three-fourths of its capacity. This provides the proper air or head space needed by the laboratory to conduct the analysis. Cut carpet samples into pieces that fit evenly into the container. The sample should be loosely rolled and placed on end into the container. Break up other debris samples to increase their surface area and place on end into the container allowing for an air space. For concrete surfaces, the suspected accelerant can be absorbed from cracks and craters with a fine powder of calcium carbonate or agricultural lime. A ⅛-inch (3 mm) layer is spread over the surface to be sampled and allowed to sit for approximately 30 minutes. The powder is then collected and placed into a vapor-tight container. A ½-inch (13 mm) layer of the concrete from the area should also be chipped up and added to the same container. Soil samples that are suspected of containing accelerant evidence should be collected using a small clean shovel or similar tool to obtain a sample of the material from the surface down approximately 6 to 8 inches (150 to 200 mm).

Accelerant samples may also be found in unburned portions of containers or liquid pools. After documentation, the fire investigator should collect a small quantity of the suspect liquid in a sealable glass vial (Teflon®-lined cap). If there is water below the liquid, collect from both layers. Mark the vial and securely pack into a small metal paint can. Clean white copier paper can be used to pack the vial into the can and prevent breakage. Place the remaining liquid into a sealable can, which is preserved by the investigator. If there is a container involved, seal openings, such as spouts and fill holes, with a cap or stopper. Cork stoppers work very well for this purpose because they can be easily cut and shaped to fit any size opening. The container should be carefully handled to prevent the destruction of any fingerprint evidence that could be on the surface.

Seal containers that appear to be empty. Package them for analysis because sufficient residue of the contents may remain on the interior surface for identification. For devices such as Molotov cocktails, package the container and any available fragments of glass in an airtight can and submit (Figure 10.8). As with all evidence, carefully document the item and the location in which it was discovered.

The collection of gases for analysis is not a task that the fire investigator will encounter at the typical fire scene. If gaseous samples must be taken, the most effective method is to use a gas sampling device that draws a sample of the material over a charcoal filter.

### Comparison Samples

The collection of comparison samples for laboratory analysis may be an element of collecting ignitable liquid evidence. During the combustion process, some materials will give off volatile hydrocarbons. Building components such as carpeting, wood, plastics, and roofing materials are examples of these materials. The comparison sample should be one of the same type as that suspected of having an ignitable liquid on it, but uncontaminated. Obtain the sample from an unexposed area well away from the location of the primary sample. Comparison samples are packaged in the same fashion as the primary sample and marked as a comparison sample. Document the location from which the sample was removed. Where there are multiple points of origin suspected, with varying types of materials sampled, a comparison sample of each type of material should be collected.

When collecting accelerant evidence, the fire investigator should make certain to perform the following:

- Take precautions during the collection, packaging, and handling to prevent contamination or spoliation of the sample.

- Take samples sufficient in size or quantity for the laboratory to conduct its analysis.

- Follow policies and procedures of the jurisdiction and the forensic laboratory that will conduct the analysis.

- Properly preserve the sample to prevent evaporation of ignitable liquids.

- Collect a comparison sample where necessary.

**Figure 10.8** The entire remains of a Molotov cocktail should be collected.

- Document all items collected before and during the collection process.

- Maintain the chain of custody from the time the item is discovered.

## Impressions

Impressions from shoes, tires, or tools are frequently found at the scene of an intentionally set fire. This evidence is very fragile and is easily destroyed during fire suppression and related activities. Scene security and early recognition that this type of evidence is present will assist in the preservation of this information.

### Collection and Preservation

For foot or tire prints made on the ground, the fire investigator will have to collect the evidence by making an impression of the print. The following steps should be followed:

Step 1: Place a ruler next to the print and photograph the undisturbed print. The camera should be directly over the print at a 90° angle. Adjust the flash so that the light does not wash out the detail of the print.

Step 2: After the print is photographed, the investigator should make a plaster or other casting material impression of the print.

Step 3: When dry, scratch the initials of the investigator and a directional mark indicating north into the back of the cast.

Step 4: Wrap the dry cast and place it into a well-padded container.

Step 5: Mark and seal the container as evidence.

Before being collected or otherwise disturbed, any tool marks found at the scene that are related to the incident should be documented. The investigator should note that the fire department may have forced openings as part of its suppression activities. Where marks are determined to be of evidentiary value, they should be collected. If possible, objects with tool marks found at the fire scene should be submitted to the laboratory intact. If the object the mark is on is too large to collect intact, the portion of the object with the mark on it should be removed, marked as evidence, and packaged. Where the mark cannot be collected, a cast should be made as previously described for foot and tire prints.

## Fingerprint Evidence

Fingerprint evidence may be found at the fire scene on a variety of surfaces and materials. This evidence may even be found on surfaces that are heat damaged or soot covered. Special processing by the forensic laboratory may be required to locate and document the prints so that they can be used.

### Collection

Materials collected for submission as possible fingerprint evidence should be collected and packaged so that the surface with the potential print is protected from any abrasion or rubbing. The evidence is fragile and can easily be destroyed if not carefully managed.

 **Conclusion**

The collection and handling of evidence is an essential part of every fire investigator's job. While performing this function, it is very important to always be alert to prevent the destruction or loss of critical evidence. Proper collection and preservation of materials at the fire scene is the beginning of the process. Avoid contamination during collection. Package materials so that they are preserved during transport. The final step is maintaining the chain of custody for all evidence collected.

Of equal importance is the necessity to protect evidence from spoliation. Whenever the evidence is to be examined by specialists, consideration should be given to other interested parties so that they may be involved in future testing or analysis.

The identification of potential evidence at the fire scene will be discussed in Chapters 11-14. The selection of laboratories and the type of analysis available from the forensic laboratory will be discussed in Chapters 15 and 16.

# Chapter 11

# Interviewing and Interrogation

## Performance Objectives

This chapter provides information that addresses performance objectives described in NFPA 1033, *Standard for Professional Qualifications for Fire Investigator*, particularly those referenced in the following sections:

**Chapter 3 Fire Investigator**
**3-5.1 a and b**
**3-5.2 a and b**
**3-5.3 a and b**

One of the most important skills the fire investigator can develop is the ability to plan and conduct a good interview. During the course investigation, the investigator uses this skill to develop information, verify opinions, and fill gaps in the information he has developed regarding a fire and its cause. Information collected during interviews may be extremely valuable to the fire investigation.

Interviews will be conducted at various times during the investigation, but the objective will always be to develop information related to the incident. Some investigators find it helpful to interview witnesses before performing the scene examination. Others wait until after they have viewed the scene before they talk to anyone. A reason for this is that the information gathered during this process could be helpful in developing questions to ask during the interviews. The investigator uses his judgment and experience to determine the timing of interviews. However, the fire investigator should not delay interviews. Witnesses at the fire scene with valuable information may leave the area without being identified, or they may be unavailable at a later time.

This chapter will primarily be devoted to the methods used for developing information through interviews. It is, however, important for the fire investigator to understand the difference between an interview and an interrogation. An *interview* is the questioning of an individual for the purpose of obtaining information related to the investigation. Interviews are nonaccusatory in nature. An *interrogation* is a formal line of questioning of an individual who is suspected of committing a crime or who may be reluctant to provide answers to the investigator's questions. Interrogations are accusatory in nature; that is, the individual being questioned is a suspect in the commission of a crime such as arson.

If during the course of an interview, the line of questioning changes from nonaccusatory to accusatory, the fire investigator must be cautious not to violate the rights of the individual being questioned. While fire investigators are primarily involved in conducting interviews rather than interrogations, it is very important that they know and understand the differences and comply with the law as it relates to interrogations. If the person is taken into custody or deprived of freedom in any way, the person must be advised of his constitutional rights before questioning. (See the following sidebar on *Miranda v. Arizona.*)

# ◆ Planning the Interview

To properly plan the interviews that will be conducted during a fire investigation, the fire investigator must establish the goals for the interviews. These goals include:

- Gathering pertinent and accurate information

- Obtaining information that corroborates or refutes the investigative data available

- Assisting in the determination of the fire origin and cause and affixing responsibility

When the goals are established, the fire investigator must then identify who is to be interviewed as part of the investigation. During the initial investigation at the fire scene, police officers, firefighters, witnesses/suspects, owner (s), and occupants will be interviewed. (**NOTE:** See the end of this chapter for a list of interview questions.)

---

## Legal Issues Related to Interrogations

The Fifth Amendment to the U. S. Constitution guarantees that no person accused of a criminal offense can be compelled to be a "...witness against himself..." The Sixth Amendment goes on to guarantee the right to be represented by legal counsel when accused of a crime.

Once an individual is taken into custody or deprived of freedom (or has the perception that his freedoms are taken away), the investigator is obligated to make the individual aware of his constitutional rights. The Supreme Court's landmark decision regarding this is *Miranda v. Arizona*, 384 U.S. 436 (1965). Based on the Miranda ruling, a person in custody must be given the following warnings:

- He has the right to remain silent.

- Any statement that he makes may be used as evidence against him.

- He has the right to the presence of an attorney.

- If he cannot afford an attorney, one will be appointed for him before any questioning, if he so desires.

Any evidence or statement obtained prior to or without the Miranda warning will not be admissible in criminal court. While the actual "Miranda Warning" language may vary from state to state, the actual warning will normally be as follows:

- You have the right to remain silent and refuse to answer questions.

- Anything you say may be used against you in a court of law.

- You have the right to consult an attorney before speaking to police and to have an attorney present during any questioning or in the future at any stages of the proceedings.

- If you cannot afford to hire an attorney, one will be provided for you without cost at any time.

- If you do not have an attorney available, you have the right to remain silent until you have had an opportunity to consult with one.

- If you are willing to answer questions without an attorney, you have the right at any time to stop and consult an attorney or to stop for any reason.

- Finally, now that I have advised you of your rights, are you willing to answer questions without an attorney present?

The investigator should then ask:

- Do you understand each of these rights I have explained to you?

- Do you wish to remain silent?

- Do you wish to give up the right to speak to an attorney and have him present during questioning?

Miranda warnings are not required under the following conditions:

- When the person being questioned is not in custody —his freedom of action is not restricted in any way, and there is no perception that he is in custody.

- When the information is spontaneous.

- When being questioned by a private investigator; that is, as long as the private investigator is not working for law enforcement when the questioning occurs.

- When the questions are routine and not accusatory in nature.

Because of the implications, any interrogation of an individual must be well planned and documented. Failures to follow proper procedures under these circumstances will often result in the exclusion of crucial information in criminal cases.

Investigators should always follow local policy established for conducting interrogations. They should also use extreme caution when questioning individuals who do not speak English or have diminished mental capacities. In any special case, legal guidance should be requested before questioning.

---

After the parties to be interviewed are identified, they should be separated from spectators, firefighters, and others at the scene and from each other. Experience has shown that if witnesses are allowed to talk to each other and compare their observations, some may adjust their stories to agree with other witnesses' stories (Figure 11.1). The fire investigator evaluates all data currently available. This information is then used to assist in the development of lines of questioning for the interviewees. Specific methods and questions for the type of interview being conducted are discussed later in this chapter. If time permits and the information is available before the interviews, the investigator may review any background information on witnesses, the owner/occupants, or the reporting party. At the scene, background information might come from a police officer familiar with the location or firefighters who have previously responded to the location.

The fire investigator should then decide on the setting for the interviews. The basic considerations related to the setting include:

- Location where the investigator and interviewee are safe

- Zone of privacy for the interview

- Minimal distractions during the interview

While at the scene of an incident, the interview location could be as simple as an area away from onlookers and emergency personnel, such as in the investigator's vehicle or in a nearby home or business when permission can be obtained to do so (Figure 11.2).

As the investigation progresses, additional interviews may be necessary. The investigator can more easily control the setting for these sessions. He can conduct them in an interview or conference room at the fire or police department or in the interviewee's home or office. The investigator can minimize distractions by making an appointment and making it clear that a period of uninterrupted time is needed. For follow-up interviews, the investigator should consider whether prior notice would help or detract from the information that will be obtained. Giving notice to a suspect or an unwilling witness might not be advisable. On the other hand, follow-up interviews with bankers, insurance agents, or other disinterested parties might require prior notification.

**Figure 11.1** Witnesses allowed to talk with each other may compare their observations and adjust their stories.

**Figure 11.2** The interview location at the scene of an incident can be as simple as an area away from onlookers and personnel.

Final considerations related to planning an interview are to determine who should be present during the interview and who should conduct the interview. It is normally best to interview one individual at a time. If the individual being interviewed is the opposite sex or if there is some concern for the investigator's safety, another investigator, police officer, or member of the fire department may be asked to witness the interview.

## ◆ The Interview

Once the investigator has developed a plan of action and a list of people to interview, he can begin the process. The investigator conducting the interviews should always keep in mind that the purpose of the interview is to collect useful and accurate information about the fire and its cause.

The first step in the interview should be introductions. The investigator should introduce himself and provide interviewees with identification. The investigator can set the tone of the interview during this early stage. Depending on the type of interview and the person being interviewed, the approach can be casual, friendly, businesslike, or authoritative.

Investigators select their approach for the interview based on their experience and what works best for them. They will generally modify their approach depending on the type of interview being conducted. Interviews with firefighters, police officers, and others without a specific interest in the outcome of the investigation can be approached with an attitude of trust. Investigators should use more caution while interviewing people with an interest in the outcome of the investigation, such as the property owner or occupants of a fire building. During the initial interview, the investigator must determine whether the information provided is reliable. When the investigator interviews an individual who is a suspect in an incendiary fire, he must regard the information provided with extreme caution.

Before asking any questions, the investigator should positively identify the person being interviewed. The investigator should use a form, such as NFPA 906-6 (see Appendix C) or a similar form developed by the jurisdiction, to record the following information for each person interviewed:

- Full name
- Social security number (This is optional and cannot be required)
- Date of birth
- Home address
- Home telephone number
- Work address
- Work telephone number
- Physical description (optional)

The investigator should also include the date, time, and location of the interview and the name of the investigator conducting the interview.

Once the formalities of introductions and identification are completed, the investigator should begin the interview by outlining the objectives of the interview. During this opening, the investigator should be careful to disclose as little information as possible about the incident. *It is important to remember that the objective is to get the person being interviewed to give the investigator information regarding the incident.*

During the interview, the investigator should be positive and professional. It is important to be objective and focus on getting as much information as possible. The most effective approach is to ask open-ended questions and allow the interviewee to speak freely without interruption. By using this approach, the interviewee will usually give the investigator more information than if very specific, short-answer questions are asked. The investigator should not interrupt while an answer is being given. Interruptions may make interviewees uneasy and cause them to stop talking. Direct questions can be asked after an interviewee has completed his story or if he is reluctant to talk to the interviewer. The investigator should phrase direct questions so that the answers provide the information he is seeking.

While the interviewee is responding to questions, the investigator should actively listen while making eye contact. Active listening is a skill that can be learned. The investigator should use the following skills for active listening:

- Maintain eye contact with the interviewee. Listen with your eyes as well as your ears.
- Let the interviewee know that you are paying attention to what is being said by nodding or making attentive noises.
- Stay focused and involved.
- Avoid emotional involvement. Remain objective and open-minded.
- Do not let your mind wander while the interviewee is speaking.
- Concentrate on what is being said.
- Ask yourself mental questions. For example, how does what is being said apply to the investigation?
- Do not let your mind wander during the gap between what is being said and your thought process. Anticipate what the interviewee is about to say. Use this time to make notes or to develop additional questions for the interviewee.

The investigator should note any discrepancies in the interviewee's statement and also be prepared for spontaneous statements. The investigator should avoid being caught off guard or reacting to the information that is being provided. Based on the information the interviewee provided and the knowledge the investigator has of the incident, the investigator should be prepared to vary the line of questioning in order to gather additional information the interviewee might be willing to provide.

The two most common mistakes that interviewers make are talking too much during the interview and not allowing the interviewee to answer questions or tell his story. The interviewee should not walk away from the interview knowing more about the incident than was known prior to the interview. The investigator should say as little as possible, attempt to put the interviewee at ease, and allow the interviewee to do the majority of the talking during the interview. Some investigators will let the interviewee tell the story several times before asking specific questions.

As the interview progresses, the investigator should always be alert to nonverbal indicators that the interviewee is under stress or acting in a deceptive mode. Stress indicators the investigator should look for include depression, denial, bargaining, and acceptance (absence of fight-or-flight).

Nonverbal indicators are often called *body language*. Any of the following listed responses may indicate that the witness is stressed by the interview and this may or may not suggest that he is being untruthful.

- Head position
- Facial color
- Expressions
- Covering or blocking mouth
- Scratching nose
- Touching head
- Yawning
- Biting nails
- Cleaning fingernails
- Playing with hand or rings
- Exhibitions of anger
- Defensive postures

- Breaking eye contact
- Sitting postures
- Chairs as barriers

At the close of the interview, the investigator should inform the interviewee that the information provided will only be used as part of the investigation and that it is considered confidential. The investigator should give the interviewee a chance to provide any additional information before ending the interview. Many times, information obtained at this point will be helpful to the investigation. It is given in a less formal atmosphere, and the interviewee may have dropped his guard thinking the interview is over.

## Documenting the Interview

Like other parts of the investigation, the investigator must document each interview. The simplest and most common method of documentation is note taking. In this case, the investigator obtains the basic information regarding the interviewee as previously discussed and then makes notes during the interview. However, this method can be somewhat disruptive and could result in information not being obtained due to the distraction or the investigator not hearing the information because he was concentrating on writing the information.

Where it is permissible with the jurisdiction, and the interviewee grants permission, making an audiotape or videotape of the interview is an excellent method of documentation (Figure 11.3). By using a recording method, the majority of the discussion is obtained, and the investigator can pay closer attention to what is being said rather than taking notes. The tape can also be transcribed at a later date and

**Figure 11.3** An audiotape of the interview is an excellent method of documentation.

used during the analysis portion of the investigation or at trial.

Several methods may be used during the interview to obtain a statement. These include:

- The interviewee can write and sign the statement.

- The statement can be dictated, transcribed, and signed.

- Audio or video recording can be used to document the statement.

All statements should include the date, time, person making the statement, and individuals present during the statement.

 ## The Evaluation of Interview Information

As soon as the scene examination and related interviews are completed, the fire investigator should organize and analyze all information collected. This analysis will identify any gaps in the investigative data available, and it can be used to develop a plan for additional interviews or data gathering. The investigator should review the interview information and identify the data that supports the findings. Also, it is important to identify areas for further investigation where interview information conflicts with information from the scene or other interviews.

During this analysis, the fire investigator evaluates the information provided by the various individuals interviewed and judges the reliability of that information. An example of this type of judgment would be descriptions of the fire from occupants, witnesses, and firefighters. Based on their experience and knowledge, the description provided by the firefighter would probably be the most accurate. Information provided by parties who are suspect in criminal investigations should be evaluated with caution. To assist in this evaluation, the investigator might use an index to rank the validity and reliability of the information available.

As the interview information is organized and evaluated, the investigator can begin to develop hypotheses or conclusions based on the facts that emerged during the investigation.

 ## Conclusion

Developing information based on interviews is an important part of the fire investigator's job. The fire investigator must develop skills in order to collect accurate information, detect inaccurate information, and accurately analyze this information along with the other investigative data. (**NOTE:** See Figure 11.4 for topics to cover and questions to use when interviewing police officers, firefighters, witnesses and/or suspects, the owner, and occupant/insured.) The investigator must always be aware of the legal issues surrounding interviews and interrogations and take measures to ensure that any information developed is done so is a legal manner.

# Topics to Cover When Interviewing Police Officers

## Alarm Information

Describe the alarm.

Can you identify the reporting party or parties?

What was the time of the alarm?

Do you have any other comments regarding the alarm?

## Arrival at the Scene

What was your time of arrival?

What were your observations during travel to the fire scene?

Provide a general description of property upon your arrival.

Describe the fire and/or smoke.

Did you notice anything unusual with the structure?

Were persons/vehicles arriving or leaving the fire scene?

Did you take any statements or hear any comments by persons at the fire scene?

Did you recognize anyone who is common to fire or other emergency incidents?

Were there any distractions upon arrival or during scene security?

## Activities at the Scene

Did you arrive before or after the fire department?

Were there any rescues or attempted rescues during the incident?

Were the doors and windows opened or closed, locked or unlocked?

Did an officer conduct any activities? If so, explain.

Were there area(s) where the fire appeared to be more intense?

Describe the area(s) where the fire had burned through the exterior walls or roof.

Did the police or fire department secure the scene? If so, explain.

Did you conduct canvas neighborhood interviews?

What is your opinion as to the area of origin?

Did you observe any unusual odors?

What were your observations of the flame spread and intensity?

Did you notice any vehicles that passed the scene more than one time? If so, did you observe any unusual actions by person(s) in the vehicle?

Were any arrests made at or near the scene?

Did you receive any domestic disturbance calls at or near this fire scene?

Do you have any comments about items not mentioned?

## Other Observations

Did firefighters or police officers make any comments?

Is there a history of police calls to this location?

Did the police have concerns regarding this neighborhood?

**Figure 11.4** Interview questions/topics.

## Topics to Cover When Interviewing Firefighters

### Alarm information

What method (9-1-1 or walk-in) was used to report this alarm?

Can you identify the reporting party or parties?

What was the time of the alarm?

Did you notice any normal or unusual situations regarding the alarm received?

### Arrival at the Scene

What was your time of arrival?

What were your observations during travel to the fire scene?

Describe the fire and/or smoke.

Did you notice any burn-throughs of the exterior walls or roof?

Were there any hydrant problems?

Describe any problems with the fire department connections or private fire system, if applicable.

Was there fire alarm/smoke detector activation?

Did you notice anything unusual with the structure?

Upon arrival, were persons/vehicles at the scene? If so, were they arriving or leaving?

Did you take any statements or hear any comments by persons at the fire scene?

### Suppression Activities and Observations

Were the doors and windows opened or closed, locked or unlocked?

Who were the first-in firefighters? What were their observations?

Explain the activities during the fire incident.

Was your department using an offensive or defensive attack?

Describe areas where attack lines were used.

What size attack lines were used?

Did flashover occur?

Was there any evidence of an explosion?

Were there any rescues or attempted rescues during the incident?

Describe the location and types of ventilation used.

Were usual or unusual suppression activities used?

Which area(s) are the most difficult to extinguish?

Were any rekindles experienced?

Did you observe any low-level flames?

Do you have an opinion as to the area of origin?

Did you observe any unusual odors?

Did you observe any unusual flame spread or intensity?

Were there any unusual observations of the fire/smoke?

Describe the conditions of the utilities at the fire scene.

Do you have any comments about items not mentioned?

Figure 11.4 continued.

## Other Observations

Did the witnesses and/or firefighters make any comments?

Did the owner(s) or occupants(s) make any comments?

Did the owner or occupant sign a consent form?

Did you notice persons who have been seen at other fire occurrences?

Did you find or seize any physical evidence?

### Interview with Witnesses and/ or Suspect(s)

Before the start of the interview, the investigator should record (in written format, audiotape, or videotape) the following information:

Date/time of interview

Location of interview

Case number

Tape/side number

Date of fire

Location of fire

Acknowledgment of recording device

Status of the interviewee: owner, renter, neighbor, etc.

Persons present at interview

### Ask interviewee to:

State full name, and spell each.

State date of birth.

State social security number. (Optional — cannot be required)

State current address and length of occupancy.

State phone number — home/work/cellular/other.

The investigator reads *Miranda* if applicable.

### Begin interview with the following questions:

Are you under the influence of any medication at this time?

Are you under the influence of alcohol or other substance that may affect your answers?

Do you have tattoos or other obvious markings? If so, identify them.

Do you have an arrest history?

Do you use any aliases or nicknames?

Have you ever been investigated?

How long have you lived in this city?

Where were you at the time of the fire?

Have you ever started a fire?

Figure 11.4 continued.

Was anything moved before the fire?

How do you think the fire started?

How did you find out about the fire?

Is there a history of fire occurrence at this location?

What were your actions upon finding out about the fire?

Who do you think had the best opportunity to set this fire?

**Continue interview with additional questions:**

Did you like this house or business?

Who sold the structure to you?

What agencies were trying to sell the structure for you?

Who helped you move the contents out of the structure?

Were the electricity/gas/water turned on at the time of the fire?

Were any appliances on at the time of the fire?

Have you been inside the house after the fire?

Was anything missing from the structure?

Identify and give locations for any personal pictures inside the structure.

Identify and give locations for guns inside the structure.

Were there any problems with utilities, appliances, or other items?

Had the utilities, appliances, or other ignition source items been repaired?

Do you have conflict with neighbors or family?

Where was the last place you purchased your gasoline?

Did you talk to anyone before talking with me?

Have there been any repairs/renovations to the house (business)?

Provide financial account location and numbers. Include mortgage history and information on property (ies).

What is the monthly income from employment? What is the total family income?

Have you ever been late with payment of expenses?

What was your financial condition prior to the fire?

What do you consider your financial condition to be after the fire?

Have you contacted your insurance company?

What do you think should happen to person who set this fire (if applicable)?

Who was the last person inside the house or business?

Were normal procedures used for closing or securing of house (business)?

Were doors/windows locked at time of fire?

Who did you ask to set this fire for you?

Figure 11.4 continued.

Is there any reason why someone would tell me that they saw you at the house (business) just before the fire?

Tell me the reason why your fingerprints were found at the scene.

Have you answered all my questions truthfully?

Will you agree to a polygraph test?

How do you think you (will) would do on a polygraph test?

Who do you think set this fire?

Did you set this fire?

Would you be willing to make restitution just to get this over with? (If yes, ask person to make restitution to a similar but unrelated incident — validation.

Have you made a list of items lost in the fire for the insurance company?

Where do you usually buy gasoline (diesel) for your vehicle(s)?

Will you vouch for anyone you feel did not set this fire?

## Research the following information:
History of burglary claims from insurance companies

Insurance coverage/building content and additional living expenses

Replacement cost for loss

Any changes in the policy

Agent and insurance identity

History of entire family regarding fire occurrences

History of residency

Educational history

History of service in armed forces

Marriage and family history

Employment history

## Other investigative topics
Allow interviewee to clarify or comment on items not questioned on. Follow up with questions relating to answers that appear deceptive.

Change chronological order of activities of suspects.

### Owner Interview — Financial Aspects

What type of business was involved in the fire?

How long have you been in business?

Who did you buy the business from and what were the terms of the sale? (Assume that the property is owned, but if it is leased, get data on terms of lease.)

What was the form of business organization — sole proprietor, partnership, or corporation?

Figure 11.4 continued.

Are any other locations used in conjunction with your business for storage operations?

Do you have any other business? What type?

What was your initial investment in the business and have you made any additional deposits to support the business operation?

What type of products/services do you sell?

Has the product mix of your sales changed recently (for example, from the top of the line products to the cheaper discount goods)?

Have you made any recent repairs or renovations to the property?

Did an outside service contractor make these repairs/renovations? (If so, obtain details of contract.)

Are you planning to rebuild?

Was the business for sale and, if so, how much?

If you were to sell the business, what would be the estimated selling price?

Do you have an accountant? (If so, obtain details.)

What services does the accountant provide?

Have you filed federal and state income tax returns for the past five years?

Who prepared these returns? Do you or your accountant have copies?

Did your accountant prepare financial statements? If so, for what years and are they available?

What was the financial condition of the business at the time of the fire?

Was the financial condition of the business improving or declining at the time of the fire?

What were the average monthly gross receipts? Expenses? Net income?

Were there seasonal fluctuations in your business?

Were you paying all your bills in a timely manner?

Did any suppliers put you on COD?

Did you maintain an adequate supply of inventory items to support your sales?

Did you have any old and/or obsolete inventory? What products were involved and what was the approximate inventory value?

Please provide us with a complete list of all your main suppliers and contractors.

Have any judgments, liens, or lawsuits been filed against you or your business?

Have you ever filed or are you in the process of filing for bankruptcy?

Did you draw any type of a salary from the business?

Do you have any other sources of income? (If yes, explain.)

Is your spouse employed? (If yes, obtain particulars as to salary, employer, etc.)

Do you have a business checking account? (If yes, obtain account number, bank etc.)

Do you have any other accounts?

Are all business deposits and disbursements made to these accounts?

Do you ever commingle any of your personal funds with the business accounts?

Have you ever loaned any of your personal funds to your business and if so, how much, when, etc.?

Figure 11.4 continued.

Do you have any outstanding loans, mortgages, etc.? (Obtain complete loan data as to amounts, terms, banks, etc.)

Has any lending institution ever turned down any of your loan requests?

Are any of your loans in default?

Do you have any automobile loans not previously mentioned?

Do you have any credit cards? (If so, obtain type of card, bank, account number, and outstanding balance.)

Have you been paying all your obligations in a timely manner?

Have suppliers been paid in a timely manner? If not, why and what is the outstanding balance?

How does the total amount that you currently owe to suppliers compare to this same time last year? Two years ago?

Do any of your customers owe you a significant amount of money? If so, who are they, and why have you not been paid?

## Fire Loss Questions for Occupant/Insured

Date

Name

Address

Investigator

Time

File #

Please answer the following questions regarding your fire loss. This information will assist in the investigation as to the origin and cause of the fire.


Were you or any member of the household present when the loss occurred? If yes, where was the fire first observed?

What time did the last person leave the house?

Who was the last one to leave the house?

Were the doors and windows locked and secured?

Who had keys to the property?

What were the weather conditions at the time of the loss?

Were there any problems with the gas or electric utilities or appliances? If yes, describe.

Have you purchased or received any new appliances within 90 days before this loss? If yes, please list appliance(s) and place of purchase(s).

Have you had any utilities or appliances repaired within 90 days before this loss? If yes, describe repairs and list the repair person or company:

What was the primary heating system (central heat, gas, electric, or other)?

Were any auxiliary-type heating devices used (wood stoves, kerosene heaters, electric space heaters, and/or gas heaters)? If yes, please list type and location.

Figure 11.4 continued.

Do any occupants of the house smoke? If yes, where were they smoking before the fire?

Were there any flammable products, such as gasoline, kerosene, camping fuel, etc., stored inside the structure? If yes, what and where was it located?

Were there any guns in the structure? If yes, what type and where were they located?

Who discovered and reported the fire, if you know?

What time was the fire reported, if you know?

Have you ever had any fire losses? If yes, when and where?

Have you ever had any theft or vandalism losses? If yes, when and where?

Have you determined if anything is missing from the structure? If yes, what is missing and from what area?

What was burning when the fire was discovered, if you know?

Do you have any opinions as to what may have caused the fire? If yes, explain.

Were the utilities connected and turned on at the time of the loss? If no, what was off?


Occupant/insured                                    Date

Spouse/co-insured                                   Date

Figure 11.4 continued.

# Chapter 12

# The Exterior Examination

**Performance Objectives**

This chapter provides information that addresses performance objectives described in NFPA 1033, *Standard for Professional Qualifications for Fire Investigator*, particularly those referenced in the following sections:

**Chapter 3 Fire Investigator**
**3-2.2 a and b**

The preceding chapters of this manual have been devoted to the knowledge and skills the fire investigator needs in order to perform an investigation. The following chapters discuss how the knowledge and skills are applied to the investigation of a structural fire.

Performing the exterior examination at the start of a fire investigation provides the fire investigator with an overall understanding of the fire scene before entering the building (Figure 12.1). During the exterior examination, the fire investigator should make observations or determinations on the following:

- Scene safety

- Scene security

- Location and condition of potential evidence

- Fire damage on the exterior of the structure

- Evidence of fire department operations (forcible entry, overhaul, etc.)

- Location and condition of building utilities

- Location and condition of any visible building systems

- Contents removed from inside the building to the exterior

- General conditions of the building and surrounding area

As the fire investigator conducts the exterior examination, he looks for any indicators that will assist in the determination of the area of origin, the materials first ignited, and the cause of the fire. Often, indicators found on the exterior of the building give clues to fire suppression activities; the behavior of the fire, including ventilation and possible locations of interior fuel packages; and insight

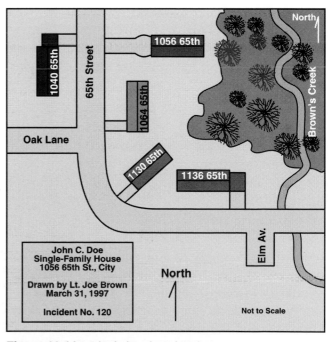

**Figure 12.1**A typical site plan sketch.

as to the area(s) of origin and evidence that may affix responsibility. The final determination of the origin and cause of the fire is based on the observations made by the fire investigator during both the exterior and interior examinations.

Fire investigators use their knowledge of scene documentation to record their findings. The investigator may record his findings by writing field notes, taking photographs, videotaping the scene, or describing the scene on audiotape. Depending on the investigation, the investigator may decide to use any or all the techniques discussed in Chapter 9, "Documenting the Scene."

 ## Conducting the Exterior Survey

During the exterior survey, the investigator walks around the fire building and the surrounding area. As previously discussed, the fire investigator assesses the safety of the building and the security of the fireground and adjacent areas. He also makes observations regarding the fire and the resulting suppression operations. This information becomes an important part of the origin and cause determination, and it may help to affix responsibility for the fire.

While conducting the exterior survey, the fire investigator uses his knowledge to determine fire behavior, interpret burn patterns, determine building construction types and systems, and to collect evidence. The following sections provide detailed information on conducting the survey and the importance of the various considerations to the overall investigation.

### Safety

As the fire investigator begins the exterior survey, he should keep in mind that even though the objective is to determine the origin and cause of the fire, his safety and the safety of those working with him are of primary concern. The investigator must wear the proper level of protective clothing and equipment to work safely in a structure that has been damaged by fire. Additionally, the investigator should be aware of applicable state and federal safety regulations (such as OSHA) related to operating in potentially hazardous environments or locations (Figure 12.2).

**Figure 12.2** Respiratory protection is vital for investigators.

Other safety-related considerations the fire investigator must make during the exterior survey include:

- Noting structural damage that could have weakened the building

- Observing any damage to utilities that could be a danger during the investigation

- Identifying any hazardous materials on the property or near the building that could pose a danger to the investigators

- Identifying the presence of guard dogs or other dangerous animals on the property

- Noting any potential hazards such as blocked egress, fall hazards, confined spaces, or areas that could entrap or otherwise endanger investigators working in or around the building (Figure 12.3)

While the investigator continues the exterior survey, he should determine the building construction type. Chapter 3, "The Basics of Building Construction as it Relates to the Investigator," describes the five major classification types as defined by NFPA 220. The investigator should note the type of materials used for the roof, the combustible materials used for the exterior finish, and the potential avenues for fire travel due to the con-

**Figure 12.3** An uncovered well can be a serious exterior hazard. *Courtesy of Bobby Henry.*

**Figure 12.4** Investigators must assess fire damage.

struction type. The construction type may not be obvious from exterior observations alone; therefore, the investigator should determine the use or occupancy of the structure. This information could provide an indication of the type of fuels involved, potential hazards, or impact of fire growth and development. While the information regarding the building and its type of construction will be evaluated at this point for safety purposes, it will also be used to interpret burn patterns and to track the spread of the fire as the investigation progresses.

As the investigator continues the examination of the exterior of the building, he should note any visible structural damage. Knowing how the building is constructed and observing damage present, the fire investigator should be able to make a preliminary determination as to the stability of the structure. This determination will be based on fire damage to structural members, deformed steel beams, or fire damage to wood truss floor or roof members. Any of these conditions are indicators that the structure could be dangerous for investigators to work in or around. Any indications of potential structural collapse in the building should also be taken into consideration. The investigator should note the areas of the building where there was intense burning and should inspect floors, where possible, for a burn-through (Figure 12.4).

The investigator should also look for indications of damage or operations that might have weakened the structure during suppression. This could include the removal of structural components or the application of large quantities of water into the building. Keep in mind that every gallon of standing water in a building weighs 8.33 pounds (3.7 kg). This weight, added to an already weakened structure, presents a significant collapse potential.

Before entering the structure, the fire investigator should examine all means of egress from the building. The investigator should note any potential hazards such as fire damage to exit stairs, doors blocked by debris, or holes burned in floors near the exits. Knowledge of usable exits is important while working in the structure.

There are many ways to make a scene safe. Techniques such as shoring, water removal, securing utilities, and other hazard abatement methods should be used as necessary. The fire investigator may require qualified outside assistance to make the scene safe for entry. The ultimate goal is the safety of the personnel working at the scene. Scenes that pose undue risk to investigators should not be entered.

## Scene Security

The fire investigator will assess the security of the fire building and determine whether additional measures should be taken during the exterior survey. Chapter 8, "Securing the Fire Scene," provides information on how to secure the scene and any investigative findings. As part of the exterior examination, the fire investigator should look beyond the immediate area of the fire building for potential evidence. If new evidence is located during the survey, he will expand the security zone around the building to protect the evidence. Entrance/egress points into the security zone may need to be secured to prevent unauthorized entry.

The investigator must be confident that the scene is secure. Unauthorized entry is not allowed, and all potential evidence must be protected from further damage or possible destruction. If this is not the case, immediate steps must be taken to enhance existing security or to expand the protected zone.

## Identification of Potential Evidence

As the exterior survey is conducted, the fire investigator is always watchful for potential evidence that may help determine the origin and cause of the fire. Examples of this type of evidence include:

- Indications of forced entry through doors or windows

- Containers of ignitable liquids in the area of the building

- Indications of multiple points of ignition

- Suspicious foot or tire imprints

- Incendiary or timing devices in or near the building

- Debris in unexplained locations

- Smell of ignitable liquid

- Evidence of tampering with building utilities

- Evidence of prefire conditions, such as open windows, that could have increased the damage caused by the fire

- Evidence of conditions, such as blocked entrances, that impeded fire department response or suppression operations

- Explosion evidence, such as clean glass shards, distant from the structure

- Ignitable liquid or solid combustible trailers

Any material located on the exterior of the building that might become evidence is secured and protected from destruction. If the fire investigator is unable to protect the material in its current location, he should document the findings and take measures to collect the item or material as soon as possible. Specifics of documentation and evidence collection are found in Chapters 9 and 10 of this manual.

Contents from the interior of the building may be found outside the structure. The investigator must determine whether these were removed before, during, or after the fire. If contents were removed from the interior, the investigator must determine who removed them.

## Fire Patterns and Indicators Found During the Exterior Survey

Burn patterns are another type of evidence that the fire investigator often observes during the exterior survey. Fire damage to the exterior of a building can be caused by fire and products of combustion venting from the interior of the building or from ignition points on the outside (Figure 12.5). In this case, the venting could also result in the extension of the fire to rooms above the window that the fire vented through. As the fire investigator continues the examination of the building, this information assists in the determination of fire growth and development.

An example of a pattern that would be formed by a fire ignited on the exterior of the building is shown in Figure 12.6. In this case, the pattern indicates that the fire originated lower than the window or at the base of the structure. If there are no openings in this area, the fire investigator should look for additional indications that the fire originated on the outside of the structure. He should carefully examine all points of low burning to determine whether they are the result of drop down or a point of origin. As discussed in Chapter 2 on fire behavior, heat and products of combustion tend to rise due to convection. Pat-

**Figure 12.5** The burn pattern indicates that the fire started inside the building.

**Figure 12.6** The burn pattern indicates that the fire started outside the building.

terns indicative of low burning can lead the investigator to a point of origin. As with the first example, the resulting fire and hot gases could also result in the spread of the fire to the interior of the structure through windows on both floors.

In both examples, hot fire gases flowing over the exterior surface of the structure resulted in a visible pattern. Fire investigators use their knowledge of fire behavior and building construction to understand these patterns and to properly analyze them. As the fire grows, the hot gases produced begin to rise and form a plume. As the gases rise and get further away from the fire, the plume temperatures begin to decrease because of the entrainment of cooler air. As this cooling occurs, the gases that make up the plume begin to spread out. This spreading process results in the production of a V-pattern on any vertical surfaces with which the plume is in contact. The actual pattern is formed on the materials present on the outside of the structure such as sheathing or veneers, gutters and downspouts, wiring, shutters, and building systems. Where combustible materials are used, the materials themselves may be involved in the fire. Where noncombustible materials are used, they may be discolored or damaged by the smoke and heat. The investigator should document the type of siding or sheathing used on the exterior of the structure for use in the analysis phase.

V-patterns on outside vertical surfaces are the result of the fire plume (flames, hot smoke, and gases) flowing over or near the surface. The plume

rises as a result of the convective currents developed just above the surface of the fuel package involved. The pattern formed shows the border between the area where the plume traveled and the area outside of the plume that received less heat from the hot gases or flame. The angle that is formed by the edges of the V is dependent on several variables, including:

- Geometry of the fuel package that causes the plume

- Heat release rate of the fuel

- Effects of ventilation on the flow of the plume

- Physical characteristics of the surface on which the pattern appears, its combustibility, and how easily it ignites (Figure 12.7)

- Any horizontal surfaces, such as building overhangs, that would interrupt the flow of the fire plume

**Figure 12.7** Effects on the material involved can reveal how the fire developed.

While the width of a V-pattern found on the outside of a structure does not tell the fire investigator anything about the speed of the fire growth, it may assist in locating low points of burning that should be analyzed during the investigation. The base of the V-pattern points toward the lowest point of burning in the area where the pattern is located.

The investigator must carefully examine, document, and analyze any low points of burning discovered during the exterior examination. These patterns are frequently the result of burning debris that drops down and causes a secondary ignition; however, they may be an indicator of an outside point of origin. The fire investigator must determine whether these are distinct and separate areas of fire origin or result from the natural progression of the fire.

During the exterior survey, the investigator should be alert for the following types of patterns or indicators:

- Irregular patterns on horizontal surfaces, such as porches, steps, decks, or patios, that could indicate the presence of an ignitable liquid pool.

- Donut-shaped or ring-shaped patterns where the outside of the ring is more heavily damaged than the inside. When pools of ignitable liquids burn, the outside portion of the pool may be damaged while the inside area is somewhat protected by the cooler liquid on the surface.

- Evidence of trailers. A trailer is a narrow pattern that may result from ignitable liquids or other combustibles being placed to provide the fire a path of travel from one location to another. Trailers can be found leading into a structure, on stairs, from a doorway into a structure, or from structure to structure.

- Window glass that is broken, crazed, or stained by the fire.

As with all potential evidence that the fire investigator will use in the analysis of the origin and cause of a fire, burn patterns on the exterior of a structure should be properly examined and documented. The investigator should collect and preserve any physical evidence according to the procedures discussed in Chapter 10, "Evidence Collection and Preservation." The type of materials used in construction as well as the fire ignition and growth will determine the type of burn patterns found on the exterior.

## ◆ Fire Department Operations

During the exterior survey, the fire investigator should examine and document all evidence of fire department operations at the scene. The investigator should determine which doors and windows the fire department opened to gain access to the building or for horizontal ventilation purposes. If the fire department forced entry into the structure, these points should be identified and documented. Identifiable tool marks or boot prints on structural openings should also be documented.

Other evidence of fire department related activities that should be documented during the exterior survey include:

- Areas where overhaul operations were conducted (Figure 12.8)

**Figure 12.8** Fire department activities must be documented. *Courtesy of Joseph J. Marino.*

- Location of fire debris removed from the interior of the building and the avenue used to transport the material

- Location of ventilation openings

- The use and location of power tools

## Building Utilities

The location and condition of all utilities entering the fire building should be examined and documented during the exterior survey. Services might include:

- Electrical

- Water

- Natural gas

- LP gas

- Telephone

- Cable TV

The fire investigator should determine if the service is currently operational. If the service is off, was it operational at the time of the fire? If there is a meter on the outside of the structure, the reading may need to be documented for the investigative file. Any signs of tampering, meter removal, or meter being booted (the addition of insulators by electric utility) before fire occurrence should also be noted and documented. In the analysis stage of the investigation, this information assists the fire investigator in ruling out potential sources of ignition and may also provide indications as to whether or not the building was occupied at the time of the fire.

## Building Systems

Any components of building systems that are located on the outside of the structure should be examined and also documented. Examples of system components that might be found during the exterior survey may include fire department connections, exterior alarm signaling devices, heat pump and air-conditioning condensing units, vents, or chimneys. Components found during the survey should be carefully examined to assist in ruling out or pointing to their involvement in the ignition or spread of the fire.

 ## Conclusion

The process used for the exterior survey varies from investigator to investigator and from fire to fire. However, it must be conducted in a systematic way, and all factors discussed in this chapter must be considered. Some investigators interview available witnesses before conducting the survey, and others wait until all observations are completed. Some investigators do a quick walk around just to get an overview of the site and then conduct a slow, deliberate walking survey to carefully examine potential evidence. Other investigators prefer to do one slow, methodical exterior survey documenting and collecting evidence as they go along.

No matter which method is selected, it is important that the exterior survey be conducted to locate potential evidence, assist in the determination of the area or point of origin, and determine the overall condition of the structure before entering it.

# Chapter 13

# The Interior Examination and Determination of the Area of Origin

## Performance Objectives

This chapter provides information that addresses performance objectives described in NFPA 1033, *Standard for Professional Qualifications for Fire Investigator*, particularly those referenced in the following sections:

**Chapter 3 Fire Investigator**
**3-2.3 a and b**
**3-2.4 a and b**

Following the exterior examination of a structure involved in a fire, the fire investigator moves to the interior of the building to complete the scene examination. The interior examination is an essential step in the investigative process, particularly for incidents where the origin was within the structure or where fire spread into the structure from an outside ignition. During the interior examination, the fire investigator once again uses his knowledge of fire behavior, building construction, and other subjects, which were discussed previously in this manual, to perform the following:

- Track the spread and direction of the fire movement within the structure

- Correlate his observations with what he knows about the fire and suppression efforts

- Attempt to locate the area(s) of origin of the fire

During the interior examination, the fire investigator builds on the information that he has developed up to that point from interviews and the exterior examination.

The interior examination is the continuation of the overall scene examination conducted by the fire investigator in order to determine the origin and cause. As the fire investigator conducts the interior examination of a fire building, he should make the following observations or determinations:

- Scene safety

- Scene security

- Location and condition of potential evidence

- Fire damage on the interior of the structure burn patterns

- Evidence of fire department operations — forcible entry, overhaul, etc.

- Location and condition of potential sources of ignition such as appliances, electrical, and HVAC systems

- Contents remaining in the building

- The prefire condition of the building and contents

The interior examination gives the investigator an opportunity to:

- View the interior of the structure.

- Document the overall condition of the building after the fire.

- Identify and secure potential evidence.

- Locate those areas that warrant more detailed examination for additional evidence and the determination of the ignition source.

Using the systematic approach, the investigator continues to develop opinions regarding the origin and cause of the fire and collects information that may support these opinions or help to rule out potential scenarios.

The fire investigator should continue to document his findings by using the techniques discussed in Chapter 8, "Securing the Fire Scene." Any potential evidence identified during the examination should be secured, collected, and preserved.

 ## Conducting the Interior Examination

The examination of the interior of a fire building is conducted to determine the area and point of fire origin, the source of the heat of ignition, and the initial fuel package ignited by the heat source. The level of detail of the fire investigator's examination depends on the incident being investigated and the complexity of the structure involved. For very large or complex structures, the fire investigator determines the extent of the examination based on the incident and the information available at the time the interior examination is conducted. For example, during the investigation of a fire found to be accidental in a high-rise building, the interior examination may be focused on the floor or compartment of origin. For incidents of unknown or obvious incendiary cause, the examination may be expanded to search for evidence, including the results of unsuccessful ignition attempts, in other areas. It may also be expanded to search for materials that support the suspected ignition scenario including tools used for forced entry into the building, ignitable liquids in unexplained areas, or incendiary devices.

The detailed examination discussed in this chapter is an example of what the fire investigator would conduct while investigating a fire of unknown origin with significant damage to the structure (or portion of the structure) and its contents. Incidents where the damage is minimal — and/or the origin

and cause is readily determined — may not require an interior examination that is as detailed as the process discussed in this chapter. As always, it is the responsibility of the lead fire investigator at the scene to determine the level of detail that is required for the interior examination. This decision is based on the complexity of the incident scene and the circumstances surrounding the investigation. The senior investigator should document the process used and be prepared to explain the decisions made regarding the investigation beyond the initial determination of origin and cause.

Many investigators begin the interior examination by conducting an initial walk-through of the building and then moving to a more detailed examination of specific rooms, locations, or areas. When the fire investigator determines that the incident warrants a detailed examination of the interior of a structure, the method selected should include the methodical examination and documentation of the interior of the building, the remaining contents, and any other evidence that is available regarding the incident.

Once the investigator has completed a walk-through of the structure, he conducts the detailed examination. This detailed examination should include documentation of the following:

- Location and condition of building contents

- Location of appliances and position of controls

- Location and condition of building utilities within the structure, including the electrical panel board containing the circuit breakers or fuses for the portions of the building involved in the fire

- Other relevant information related to the incident

This detailed examination usually progresses from the area of least damage toward the area of most damage. As with the exterior examination, the fire investigator draws upon his knowledge of fire behavior, burn patterns, building construction and systems, scene documentation, and evidence collection to conduct the interior examination. The following sections provide detailed information on conducting the examination and on the importance of the various considerations to the overall investigation.

## Safety

As with the exterior examination, safety is one of the primary issues confronting the fire investigator during the interior examination. The investigator and personnel assisting in the investigation must constantly be alert for hazards to their health and safety. The investigator should consider all potential hazards that he may face during the interior examination.

Before entering the structure, the fire investigator should determine the proper level of personal protective equipment to be used in the structure during the examination, documentation, and evidence collection (Figure 13.1). As discussed previously, there are various requirements for protective equipment based on the potential hazards in the structure. These requirements may be based on the policies and procedures of the jurisdiction or other regulatory agencies, such as OSHA, in the event that hazardous materials are present in the building. Should the site be covered under OSHA or EPA regulations for hazardous materials, specific site safety plans and staff training may be necessary. It is always recommended that more than one person be present on a fire scene investigation. However, in certain hazardous locations, it may also be necessary to comply with requirements for investigators to work in teams with backup resources available on the exterior of the structure (two-in/two-out requirements found in the OSHA Respiratory Protection Standard 29 CFR Parts 1910.134 (g) and 1926). Depending on the jurisdiction, the specific occupational health and safety requirements and levels of training may differ for public and private investigators. Before entering a scene, all investigators should be aware of the regulations that apply to the jurisdiction and make provisions to meet those requirements.

The fire investigator conducting the exterior exam should perform the following specific safety-related considerations:

- Note structural damage that could have weakened the building (Figure 13.2).

- Note the condition of utilities that could be a danger while working inside the structure.

- Identify any hazardous materials in the building that could pose a danger to the investigators.

**Figure 13.1** Investigators must wear the type and level of PPE appropriate for the situation.

**Figure 13.2** Investigators must assess any structural damage.

- Determine the necessity to monitor the air quality in the structure during the examination.

- Note any potential hazards such as blocked egress, fall hazards, confined spaces, holes in floors, standing water, or areas that could entrap or otherwise endanger investigators working in the building.

The building must be made safe for entry before beginning the interior examination. Before entering the structure, investigators should identify the entrances and egress points and should always know a safe way out. As the examination is being conducted, fire investigators must also be very cautious during the examination. They should assess the consequences of their actions before moving debris, structural components, or contents so as not to further compromise the integrity of the structure.

## Identification of Potential Evidence

Throughout the interior examination, it is very important for investigators to recognize potential evidence so that it can be documented as found and preserved. Potential evidence may relate to determining the origin and point or cause, affixing responsibility, or being an indicator of motive. It is very important for the investigator to recognize the evidence and correctly correlate this evidence. For instance, forced-entry damage does not indicate how the fire began but may have evidentiary value and should be documented when discovered. It is important that the investigator establish the fire origin and cause before considering other potential evidence that may relate to motive or suspect development. Examples of this evidence may include the following:

- Burn materials or contents
- Indications of forced entry through doors or windows
- Containers of ignitable liquids in the building
- Burn patterns that indicate the origin and travel of the fire
- Indications of multiple points of ignition
- Appliances or other devices that could have been the source of heat of ignition of the fire
- Incendiary or timing devices in the building (Figure 13.3)
- Debris in unexplained locations
- Odor of an ignitable liquid
- Evidence of prefire conditions — such as open windows, holes in floors, open or removed doors, and contents piled or placed in such a way to increase the fuel load — that could have increased the damage caused by the fire
- Absence of contents for the building
- Evidence of conditions — such as blocked entrances — that impeded fire department response or suppression operations
- Ignitable liquid or solid combustible trailers

**Figure 13.3** Various incendiary devices may be found.

Any material that might become evidence should be secured and protected from destruction. If the fire investigator is unable to protect the material in its current location, he should document the findings and take measures to collect the item or material as soon as possible. Specifics of documentation and evidence collection are found in Chapter 9, "Documenting the Scene," and Chapter 10, "Evidence Collection and Preservation."

## ◆ Fire Patterns and Indicators Found During the Interior Examination

With the exception of the discovery of an incendiary device, the fire patterns located on the interior of a structure are often the most significant factors used by the fire investigator in the determination of the origin of a fire. The patterns found during the interior examination assist the investigator in determining the "history" of the fire from ignition to extinguishment. In many cases, they lead the investigator to the area, or areas, of origin. The patterns also provide information about how the fire traveled within the building after ignition and the fuels involved during the progression of the fire. The ability to identify and interpret the patterns found during the examination of the fire scene is an essential skill of the fire investigator. This section provides information on how various patterns are formed and what the investigator should look for.

NFPA 921 describes fire patterns as the "visible or measurable physical effects that remain after a fire." These effects are found in many different forms including:

- Burned or charred materials — including structural components and contents

- Smoke or soot deposits on the surfaces of contents or walls, ceilings, and floors

- Distorted, discolored, or melted materials

Fire patterns are formed on the interior surfaces (walls, floors, ceilings, and contents) of a structure as a result of direct flame contact or the exposure to heat generated by the fire. When heat causes a pattern, the method of transfer is by conduction, convection, or radiation (see Chapter 3, "Fire Behavior"). By their nature, fire patterns have visible boundaries or borders where the fire or products of combustion impacted a surface, leaving adjacent surfaces less affected or intact. These borders, or lines of demarcation, are what the fire investigator is searching for when looking for fire patterns (Figure 13.4).

When examining fire patterns, the fire investigator should consider the surface on which the pattern is found. The material of which the surface is made has a significant impact on what is observed after the fire. Combustible surfaces, such as wood paneling used as wall covering, may actually burn away and add to the overall fuel load available to the fire. When exposed to less energy from the fire, the damage to combustible surfaces may only cause charring or discoloration. On noncombustible surfaces, such as brick, plaster, or metals, the patterns may be present as a discoloration, spalling, melting, or distorting of the surface. The type of surface can also impact the type of pattern found. For example, rough surfaces are often more heavily damaged than smooth surfaces of the same material.

**Figure 13.4** The demarcation between the burned and unburned areas is clearly visible. *Courtesy of Tulsa (OK) Fire Department.*

The reason for this difference is that a rough surface presents more surface area for the hot gases or flame than a smooth surface. The increased surface area and the turbulence created as fire gases flow over the rough surface results in additional damage.

Each fuel package that eventually burns in a room or building generates its own fire pattern. It becomes the job of the fire investigator to evaluate the patterns found during the scene examination and to correlate them to a specific fuel package. As this is accomplished, the investigator can begin to develop his hypothesis as to the area or areas of origin of the fire. This process can be difficult if there are numerous fuel packages with resulting patterns in an area. The investigator has to determine whether the fuel package was involved in the ignition sequence or whether it was exposed and ignited after the ignition of the material during normal fire spread.

The typical fire patterns observed by the fire investigator during the interior examination include the following:

- Plume generated

- Ventilation generated

- Hot-gas layer

- Full-room involvement

- Clean burn

- Irregular patterns on floors

- Saddle burns

- Protected areas

- Arrow patterns

## Plume-Generated Patterns

Early in the development of a flaming fire, a cone-shaped plume of hot gas begins to form above the flames. If the fire is near enough to a vertical surface, such as a wall or large upright piece of furniture, the hot gases flow over the surface and result in a distinct pattern. Commonly called a *V- pattern*, it is one of the patterns often found during an interior examination.

As a fire develops adjacent to a vertical surface, the flames cause an inverted V-shaped pattern on the surface. The developing hot-gas plume above the flame generates a V- or U-shaped pattern. The actual shape of the pattern created by the plume will be determined primarily by the distance of the burning fuel package and the resulting plume from the vertical surface and by the height of the ceiling (Figure 13.5).

The closer the package is to the surface, the more V-shaped the pattern will be. Taken together, the patterns appear to have an hourglass shape (Figure 13.6). The height of the inverted V caused by the flame is determined by the actual flame height the fuel package generates. The V-pattern extends up as high as the plume reaches, in many cases to the ceiling of the space in which the pattern is found. The bottom of the V points at the location from which the flame was coming as the pattern was made.

In cases where the plume is interrupted by striking a horizontal surface below the ceiling level in a room, the pattern is disrupted. In these cases, the flame plume may strike the horizontal surface and spread out along the surface. Depending on the height of the surface, the pattern may not appear as a V (Figure 13.7).

The width of the V is a function of the size of the flame zone that creates the plume. For example, a fire in a small trash container close to a wall results in a pattern on the wall that was relatively narrow, while a sofa burning adjacent to a wall generates a wide pattern on the wall. As previously discussed, the angle made by edges of a V-pattern would commonly be between 10 and 15 degrees from the centerline of the flame to the edge of the pattern. No matter what size the fire, a fire plume that is undisturbed will usually re-

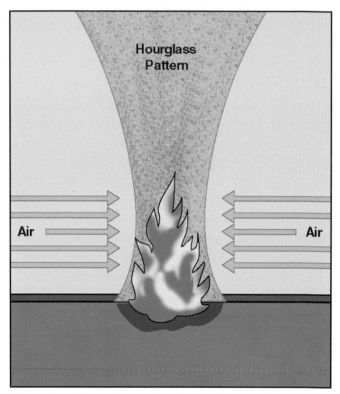

**Figure 13.6** Illustration of how the typical hourglass burn pattern develops.

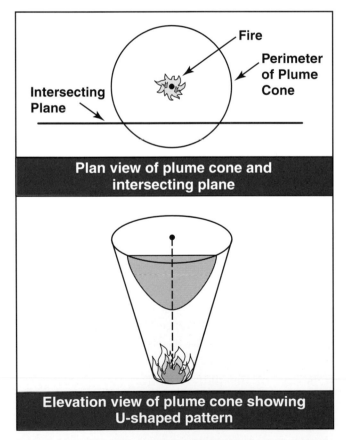

**Figure 13.5** This shows how the typical U-shaped pattern develops.

**Figure 13.7** A V-pattern disrupted by a horizontal surface. *Courtesy of Tulsa (OK) Fire Department.*

sult in a V-pattern that has boundaries in the 10- and 15-degree range from the centerline of the resulting fire pattern.

Plume-generated patterns found at the fire scene may differ from those developed under controlled research conditions due to ventilation within the space as the pattern is made. The actual shape of plume patterns is also altered when the surface that the plume hits is combustible and the surface material is ignited and burns along with the initial fuel package. Under these circumstances, the pattern generated could be very different than the "textbook" V-pattern.

The fire investigator looks for plume-generated patterns to identify potential areas of origin or the location of significant fuel packages that became involved after ignition. Plume patterns located in areas where there were no significant fuel packages may indicate that the fire was intentionally set and provide a location for additional examination and evidence collection.

## Ventilation-Generated Patterns

Many patterns found during the interior examination are the result of the movement of air in the compartment or building during the fire. The resulting patterns are present in several different configurations. The first that the investigator should consider are those that are caused by air movement over glowing embers. Cases where the embers are on a combustible floor surface could result in a hole burned in or through the floor. Once the hole is created, additional airflow in the area could result in additional damage that the investigator has to examine to determine the reason for the damage. The areas where airflow causes increased burning of embers could be confused with areas of low burning caused by an ignitable liquid pour or other intentional ignition. The fire investigator should carefully examine these patterns and collect evidence to determine whether it was in fact an area of origin or the result of ventilation.

Ventilation-generated patterns are also observed on and around doors and windows of compartments involved in fire. The first condition is when air flows into the compartment from under the door while hot fire gases flow out over the top of the door. The portion of the door facing the fire shows evidence of fire damage, or charring, at the upper portion and over the top where fire gases flowed. More than likely, the bottom of the door will be undamaged due to the cooler airflow from outside the compartment into the fire.

When the hot-gas layer fills the entire compartment, the investigator is likely to find evidence of the gases flowing from the compartment from both the top and bottom of the door opening. The fire compartment side of these doors often show damage or charring from the top to the bottom on the vertical surface as well as where the gases flowed over and under the door. Evidence of the flow may also be seen on the opposite side of the door at both the top and bottom.

When examining patterns found on a door, the fire investigator should also consider whether hot debris falling against or near the lower portion of the door contributed to the development of patterns being observed. Air movement under the door can result in increased damage should hot or glowing debris be deposited there during the fire.

The fire investigator may find other patterns or effects caused by ventilation when fire gases flow through structural openings, such as doors and windows. Any time a gas flows through a restricted opening, its velocity is increased. The increase in velocity of the hot gasses may result in damage patterns that are different from adjacent spaces or surfaces. As with any fire pattern, the fire investigator has to examine the area adjacent to the pattern and attempt to determine the factors that contributed to the development of what is observed (Figure 13.8).

**Figure 13.8** The burn pattern may reveal the venturi effect caused by ventilation. *Courtesy of G. Terry Smith.*

While examining ventilation-generated patterns in a structure, the fire investigator should keep in mind that increased ventilation in an area can result in higher rates of heat release from fuel packages, longer periods of burning, and therefore greater damage in exposed locations. Understanding these patterns and how they are created assists in correlating them with other evidence obtained at the scene.

## Hot-Gas Layer Patterns

As the hot-gas layer begins to develop in a burning compartment, radiant energy from the gases begin to cause patterns within the compartment and other areas where the gases flow. Patterns formed by radiant energy are found on exposed surfaces when compartment fires are extinguished just before flashover. These patterns may be located on horizontal surfaces, such as floors and exposed surfaces of contents located in the compartment. Nonexposed surfaces, such as the undersides of furnishings that are protected from the radiation, usually are not damaged.

When the hot gases flow over vertical surfaces, such as walls and open doors, the heat energy in the gas layer is transferred through conduction to the surface. These surfaces show the effects of the heat based on the type of material from which the surface is made. Combustible surfaces are charred or burned while noncombustible surfaces are discolored or distorted.

## Full-Room Involvement Patterns

In compartments where full-room involvement (post-flashover) takes place, the damage observed by the investigator may be extensive. These compartments usually are heavily damaged from the ceiling to floor level.

The damage takes place in a relatively short period of time but will be more extensive if the fire is allowed to burn for long periods of time or if there is a heavy fuel load in the space. The fire investigator can expect to find damage to almost every surface in the compartment. Carpeting or combustible floor coverings are heavily damaged, and there may be holes burned in the floor. Because of the intensity of the burning at all levels, the undersides of furnishings may be burned or damaged. The

amount of fuel in the compartment, how the fuel is arranged, and the available ventilation determine the extent of the destruction. Fire suppression efforts may also impact the patterns the fire investigator observes during the examination.

## Clean-Burn Patterns

Clean-burn patterns are found on noncombustible surfaces where there has been direct flame contact with or intense radiant heat on the surface. Soot or smoke deposits on the surface are burned away by the flame, leaving a demarcation between the area of flame contact or intense radiant heat and the soot- or smoke-darkened area. It would not be uncommon to find a clean-burn pattern (inverted V) indicating the flame area at the base of an hourglass pattern created by a plume on a noncombustible wall surface.

## Irregular Patterns on Floors

Fire investigators often observe irregularly shaped patterns on floor surfaces. These patterns are observed as curved or pool-shaped patterns on the surface of the floor covering. The lines of demarcation observed with these patterns are dependent on the material on which the pattern is formed. Dense materials, such as wood flooring, normally show sharper lines of demarcation than materials such as synthetic carpeting. The lines of demarcation are dependent on the type of surface and the amount of heat exposure the material receives.

Irregular patterns may be formed on floor surfaces by the following:

- Pooling of ignitable liquids (Figure 13.9)

**Figure 13.9** Ignitable liquids leave an obvious burn pattern. *Courtesy of Elk Grove Village (IL) Fire Department.*

- Hot gases flowing on the floor surface

- Arrangement of contents on the floor prior to the fire leaving protected vs. unprotected areas

- Plastic contents melting, flowing, and then burning onto the surface

- Flaming debris dropping onto the surface during the fire

The cause of the pattern has to be determined by correlating nearby fuels with the pattern and determining whether there is a plausible explanation of what is observed.

A trailer of ignitable liquid or solid fuel causes a special type of irregular pattern. *Trailers* are often used in intentionally set fires to connect remote fuel packages (combustible materials, pools of ignitable liquid, etc.) with each other. The pattern found is normally narrow and extends between burned areas. Trailers may appear to be linear in nature on horizontal surfaces. Where ignitable liquids are used to form the trailer on a porous surface, the investigator may also observe a "wicking" effect at the edges of the pattern. Wicking occurs when quantities of the ignitable liquid are absorbed by the material on which the liquid is poured, and it seeps (wicks) away from the edge of the actual pour. Wicking may appear as small fingers of damage perpendicular to the major pattern. As with other irregular patterns, the pattern observed by the investigator depends on the type and composition of the surface on which the trailer was made and the material used to form the trailer (Figure 13.10).

The donut-shaped pattern is another type of irregular pattern that is usually caused by ignitable liquid pours. This pattern is formed when a pool of liquid is ignited on a floor surface and the pooled liquid actually keeps the surface beneath it cool as it burns. In this circumstance, the pattern is formed outside the pool where the burning liquid chars the floor surface. Therefore, the lines of demarcation are at the edges of the pool while the center portion of the pool shows little or no damage (Figure 13.11).

## Saddle Burns

Saddle burns are most commonly found on top edges of floor joists (Figure 13.12). These saddle-shaped patterns are the result of fire burning downward through the floor surface above the joist. The pattern is formed by deep charring of the wooden joist and is normally very localized. This pattern is caused by intense burning on the floor surface above the joist and could result from ventilation

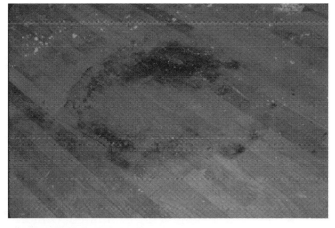

**Figure 13.11** A typical donut-shaped burn pattern. *Courtesy of Tulsa (OK) Fire Department.*

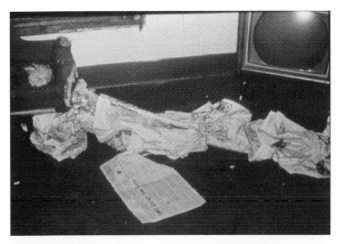

**Figure 13.10** One of several types of trailers. *Courtesy of Tulsa (OK) Fire Department.*

**Figure 13.12** A typical saddle burn. *Courtesy of Russ Chandler.*

factors, a significant fuel package in the area of the pattern, or the use of ignitable liquids. The correlation of ventilation sources or fuel packages in the area of the pattern assists the fire investigator in determining its cause.

## Protected Areas

While not actually a burn pattern, protected areas in a structure may also assist in analysis of the fire. Undamaged areas on surfaces within the fire-damaged area could have been caused by objects that shielded the surface from the effects of the fire and hot gases that damaged other portions of the surface. The investigator may find circular-shaped patterns that are the result of trash receptacles or furniture located in the area of the pattern. Bottoms of cardboard boxes that contained combustible materials, such as paper, may remain intact and protect the floor surface even though the box and its contents are burned away. During scene reconstruction, the investigator can use protected areas to place the contents back into their locations at the time of the fire (Figure 13.13). Again, the investigator will have to correlate these patterns with the fuel packages in the compartment to understand what caused them.

Examining the underside of furniture or other components in the compartment also assists the investigator in understanding the development of the fire in the compartment. Heavily charred surfaces with uncharred protected areas beneath them would be an indication that the exposure was from the area above the item being examined. Uniform charring on an item indicates heat flux from both the areas above and below the item.

**Figure 13.13** Protected areas leave a distinctive pattern. *Courtesy of Russ Chandler.*

## Arrow Patterns

When structural components, such as wood studs or trim, are exposed to flame, the sharp edges of the component are often burned away on the side of the component that faces the heat source. The result is a pointer to the heat source (Figure 13.14). If the component is part of a series, such as wall studs, the length of the remaining studs will also point to the source of the heat, with the shortest remaining stud indicating the most severe exposure (Figure 13.15). Severe charring on one side of a combustible component or surface is also an indication of that surface facing the heat source.

The fire investigator should be alert to these pointers as the examination of the fire scene is conducted. They are often helpful in tracking the source and path of travel of flame and hot gases during the fire. When reading pointer patterns, the investigator should check to make sure that the damage is fire related and not failure of breakage from suppression efforts.

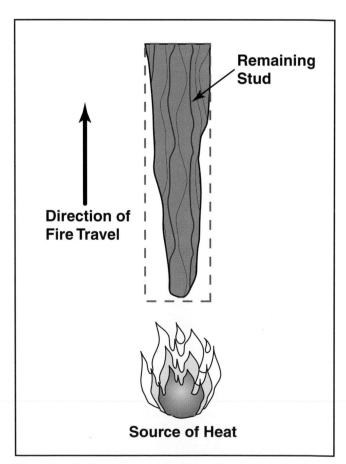

**Figure 13.14** Top view of a typical arrow burn pattern.

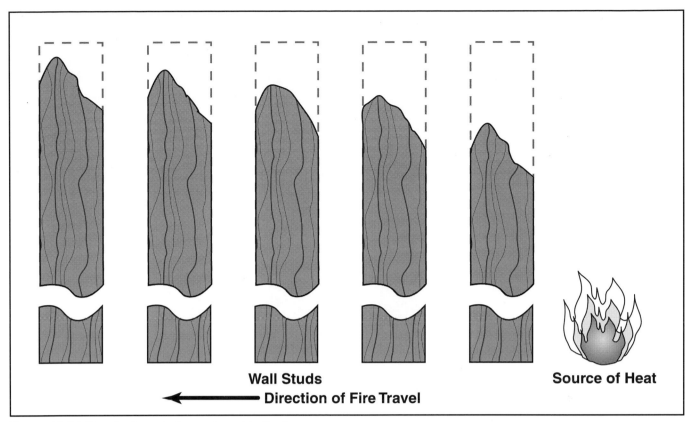

**Wall Studs**

← **Direction of Fire Travel**

**Source of Heat**

**Figure 13.15** The shortest stud was closest to the fire.

## Damage Assessment

During the interior examination, fire investigators assess the fire damage in the structure as they attempt to identify the area of fire origin. To accomplish this damage assessment, the investigator identifies, documents, and examines the patterns found in the structure. As the interior examination progresses, the fire investigator begins to identify areas within the structure that warrant additional examination. In other words, the interior examination should progress from a general examination to a more focused examination of specific portions of the structure. As the examination progresses, the investigator begins to establish a preliminary idea or scenario of what took place during the fire. Using the systematic approach or scientific method, the investigator should begin to develop a hypothesis regarding the origin and cause of the fire being investigated.

The fire investigator should identify locations within the structure that warrant additional study, and he should attempt to correlate the patterns found in these areas with the fuel packages that were located in the structure or compartment before the fire. If a pattern cannot be explained after this correlation, then additional analysis of that area is warranted. The investigator may also determine that the area is one that will be sampled for analysis by the forensic laboratory for potential evidence such as ignitable liquids. During the analysis of fire patterns, the investigator may also use protected areas to lead him to the area of lowest burning. This information, along with the analysis of adjacent patterns and the correlation with available fuel packages, can assist in the determination of the area(s) of origin.

The fire investigator should document all findings of importance during the interior examination. Typical methods for documentation have already been discussed. As part of the analysis of fire damage, the investigator may find it useful to use vector diagrams that show the direction of heat or flame spread using arrows on a floor plan of the room or structure (Figure 13.16). One method that is useful would be to use arrows to point from the areas of most damage to the area of least damage. This method should begin to direct the attention of the investigator to the sources of the most intense heat — burning fuel packages. Analysis is then required to determine whether the fuel package was

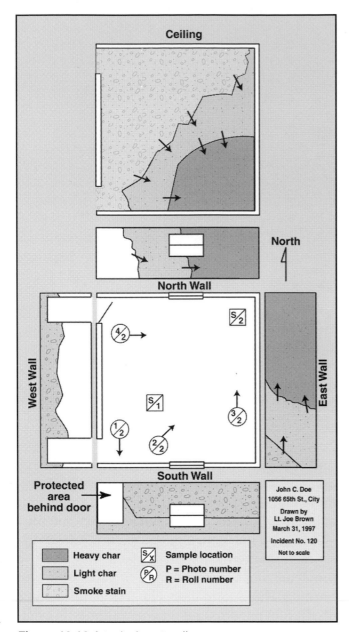

**Figure 13.16** A typical vector diagram.

correlated with prefire contents, the investigator should explore these locations to make a determination of the source of the energy.

Another documentation technique that can assist in the location of the area of ignition is the depth-of-char diagram. These diagrams are developed using the floor plan of the room and then recording the depth of char measurements made on exposed surfaces. Once the measurements are recorded, similar depths are connected using lines. The result is an *isochar* that shows intensity patterns and lines of demarcation in the area (Figure 13.18).

## ◆ Determining the Area of Origin

Determining the area where the fire began—where the ignition source and first fuel ignited came together — is an important element of determining the cause of the fire. Finding the origin of the fire involves using the fire investigator's knowledge of fire behavior, information developed from witnesses and the occupants of the structure, and the analysis of the physical information (patterns) found during the examination of the fire scene.

Once the area of origin is determined during the scene examination, the fire investigator should at-

at or near the area of origin or whether it was ignited after the fire was ignited in another location in the space.

The vector diagram is most useful when drawn on a copy of a diagram showing the prefire condition of the room or area (Figure 13.17). These drawings are developed from information obtained from the owner/occupant of the structure and should show the location of the contents — potential fuel packages—in the room or area. When the direction of heat or flame travel is overlaid on this drawing, the involvement of specific fuel packages present before the fire can be analyzed. Where there are apparent sources of heat or flame that cannot be

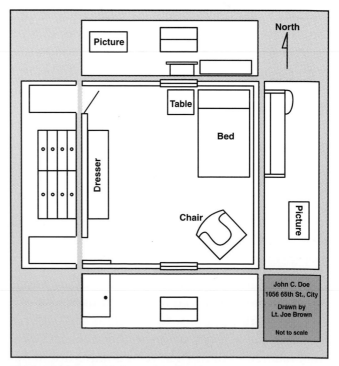

**Figure 13.17** A typical prefire diagram.

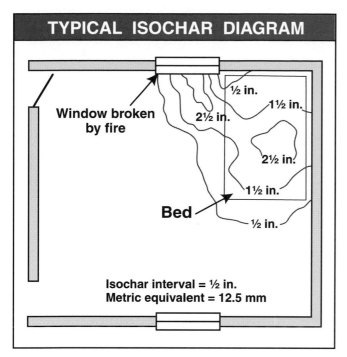

**TYPICAL ISOCHAR DIAGRAM**

Window broken by fire

½ in.

1½ in.

2½ in.

2½ in.

1½ in.

Bed

½ in.

Isochar interval = ½ in.
Metric equivalent = 12.5 mm

**Figure 13.18** A typical isochar diagram.

tempt to locate the actual point of fire origin. This is that area where the heat and fuel actually came together for the first time in the fire development sequence. Once this determination is made, the heat of ignition and the fuel first ignited can be identified. This process is normally one that involves the careful removal of debris in the area of origin using fire patterns as a guide so that the area of lowest burn or damage is located. The fire inves-

tigator will then use pointer patterns to attempt to locate the fuel package or device at the bottom of the pattern. The process used in identifying the ignition source and the first fuel involved is the topic of the next chapter in this manual.

## ◆ Conclusion

Once the examination of the exterior of a burned building is finished, the investigator must move to the interior to complete the scene examination. The interior examination is an essential step in the investigative process, particularly for incidents where the origin was within the structure or there was fire spread into the structure from an outside ignition. During the interior examination, the fire investigator once again uses his knowledge of fire behavior, building construction, and other subjects.

In the interior examination process, the fire investigator builds on the information that he has developed up to that point from interviews and the exterior examination. The interior examination is the continuation of the overall scene examination conducted to determine the origin and cause of the fire. Using the systematic approach, the investigator continues to develop opinions regarding the origin and cause of the fire and collects information that may support these opinions or help to rule out potential scenarios.

# Debris Examination, Removal, Reconstruction, and Determination of Ignition Source

**Performance Objectives**

This chapter provides information that addresses performance objectives described in NFPA 1033, *Standard for Professional Qualifications for Fire Investigator*, particularly those referenced in the following sections:

Chapter 3 Fire Investigator
3-2.6 a and b
3-2.7 a and b
3-2.8 a and b

The detailed aspects of the fire scene examination often involve the methodical examination of and the systematic removal of fire debris from the fire scene. It is during this examination and debris removal that potential evidences are identified, preserved, and documented so that they can be correctly located and further analyzed during the reconstruction phase of the investigation.

Also, once the debris has been removed, the investigator can then examine and analyze components of the building in relation to the fire. These may include building systems such as the utilities (electrical and gas), compartmentation features (fire separation assemblies and fire walls/doors), fire protection systems (detection and suppression), and HVAC systems.

 ## Debris Removal

Almost all fires result in partial to complete combustion of the fuel packages as well as structural failures. The removal of those resulting items of debris is often required for a complete and thorough scene investigation. The key objectives of debris removal are to:

- Facilitate locating the point or area of origin.
- Reveal potential ignition sources and the material first ignited.
- Reveal hidden fire growth and development patterns.
- Discover fuel packages and contents.
- Locate physical evidence.

The excavation of the fire scene is conducted in a manner similar to the tasks performed by an archeologist. For both the archeologist and the fire investigator, the layers represent periods of time. For the archeologist, the inches can represent many years, and for the fire investigator, the inches represent the history of the fire from before ignition until extinguishment. Both are looking for evidence in a form different from what it was originally. The material at which the archeologist is looking has been changed because of age and time. However, the material the fire investigator is examining may have been changed because of exposure to heat and flame and fire suppression activities. As with all operations during the scene examination, the fire investigator must proceed with caution to avoid destroying or contaminating potential evidence.

The investigator must also use caution to avoid moving debris that may affect scene safety, such as a resulting structural collapse.

The process of removal is methodical and therefore requires planning. The investigator must determine the locations to be excavated, the purpose, the layering requirements, the tools to be used, the resource requirements, and the degree of reconstruction required. Careful consideration and effort at this stage offer substantial benefit to determining, documenting, and explaining to others various factors relating to scene findings.

## Determining Locations

The investigator must determine the proper areas to first examine and excavate based on analyses of fire movement, intensity, and destruction patterns. To determine this location, the following factors come into play:

- Size of the total area damaged
- Number of rooms or compartments involved
- Damage configuration
- Amount of debris and its composition
- Patterns present to identify an origin area

It is important to remember that decisions made at the onset of debris removal often change as the process unfolds. The area exposed and reconstructed needs to be of sufficient size to provide a solid foundation for opinions and conclusions. Given these considerations, the investigator best determines the area to be excavated by carefully determining the origin area(s) from visible, evaluated, and correlated fire growth and development patterns. From this evaluation, the investigator can make the preliminary decision about where to begin the debris removal. When fire development patterns indicate a localized area, the debris removal area can be confined to a localized area. When fire development patterns fail to indicate a localized area of origin, the selected debris removal area will be proportionally larger and can encompass the entire damaged area or confines of the structural foundation. Regardless, it is important for the investigator to select an area which is large enough not only to allow for the identification of ignition sources and

first material ignited but also to allow for a thorough examination of nearby patterns and their correlation. The area should also be large enough to properly document the related patterns demonstrating initial fire development and spread.

While providing clues as to the origin area(s), damage configuration helps the investigator assess the hazards to be encountered during debris removal. It also provides information regarding what likely exists in the upper layers of the debris. For instance, broken ceiling joists or rafters indicate the potential for structural items to be discovered during the early stages of layering. *Layering* is the use of a systematic process to examine and remove fire debris. Closely related to layering are the amount and type of debris in different areas of the scene. Answering the following questions will guide the investigator toward making an initial decision about where to excavate.

- Is the debris deeper in one area than in another? If so, why?
- How consumed are the materials — ash, chunks, partial remains, or unburned?
- What visible fire development patterns correlate to the amount and type of debris?

When deciding the debris removal location, the investigator should be especially careful to maintain focus on the origin area indicators — not potential ignition sources. The investigator should not make the mistake of guessing at an ignition source and only removing debris in that area. The consequences can be inaccurate cause determinations and improperly affixed responsibility.

## Debris Removal Strategy

Once the area(s) to be excavated is established, the investigator should determine the strategy. As mentioned earlier, the layers of debris represent the time line in the history of the fire. Therefore, it is important that the debris is removed in such a manner that the sequence can be accurately determined. The debris should be examined, and then at a given area, the top layer should be removed down to the next layer and across. For example, consider finding plasterboard on the floor with the ceiling side unburned and the attic side burned. Yet, if the fire started in the room, it could be expected that the ceiling side would demonstrate greater damage and may be found in the upper levels of the debris. A layer of white ash may be indicative of more complete burning, possibly resulting from lack of extinguishment during suppression efforts.

In the immediate origin area, debris removal and layering are tedious endeavors requiring careful attention to detail. Hand tools, such as trowels, whisk brooms, and small sieves, are frequently used during this phase. If this area is substantially damaged, the tools must be used in a manner to identify and preserve small items and materials such as electrical components, metal parts from furniture, or other items potentially valuable in determining the ignition source or other material first ignited. Small sieves can be very useful for identifying these items.

If the area to be excavated or the debris is large, teams or heavy equipment can be used. Extreme care should be exercised if heavy equipment such as front-end loaders or backhoes are used. The investigator should ensure that equipment operators conserve the integrity of the structure and the scene and that they do not increase the hazards to other personnel or themselves. Of special concern is the damage or destruction that the equipment may cause, particularly to the floor. The operator must be clearly instructed to use the equipment in a manner consistent with the layering technique. Teams using hand tools complement the use of heavy equipment after each layer has been removed.

If teams are used to remove debris, the investigator should instruct them in the process of debris examination and removal as well as provide continuous supervision. It is important to keep these teams to a manageable size for purposes of examination and documentation. Regardless, the investigator must continually examine the debris being moved. If it is suspected that the fire may involve the presence of ignitable liquids, care should be taken to avoid contamination. All equipment fluids should be sampled for purpose of comparison to those samples that may be taken from the scene.

Recommendations for removing debris with a view to preserving evidence for laboratory examination can be found in the most current reference material, such as NFPA 921 and *Kirk's Fire Investigation*, and in guidelines developed by the National Institute of Justice and the National Center for Forensic Science (NCFS). The NCFS is affiliated with the University of Central Florida in Orlando and is an excellent source for information about forensic sciences, particularly fire and explosion debris analysis, and for a list of reference materials. The NCFS website is www.ucf.edu/NCFS.

For large areas or areas that are heavily damaged, establishing a grid is a useful method that promotes systematic examination, helps to document the process, and aids in proper reconstruction. When using a grid system, the floor area is subdivided into manageable segments that are marked using string, rope, barrier tape, or some other similar marking method. When the space is subdivided, the investigator can then begin the debris removal process one grid at a time. When a grid is used, the investigator should document the grid-search sequence in his notes. The investigator should also insert a photo of the grid system being used for debris removal at the fire scene. He should also provide a sketch of a grid with identification markings.

## The Layering Process and Managing Debris

As part of the planning process for this phase of the fire investigation, the investigator assigned to the task should identify the path of travel that will be used to move the material from the site being excavated. The debris should be removed from the scene to a location where it can be examined in greater detail. When moving the material, it is important to prevent contamination or destruction of potential evidence in other areas of the building.

During the removal of fire debris, the fire investigator may locate structural components, such as ceiling materials or wall-framing components. The investigator should carefully examine and document these components. If the investigator can determine the prefire location of a component, the patterns he finds on the component may assist in determining the direction of fire exposure and spread. During the layering process, the fire investigator might ask the following questions:

- What is the prefire location of the material being examined?

- Are there directional indicators on the material (did the fire come from one side or the other)?

- Was the debris layer created by the fire or as a result of fire suppression activities such as overhaul?

The investigator should take photos of each item as it was found and document its location on a floor plan or drawing being used to document the removal process. After the items have been documented, the investigator may decide that the debris or materials being removed from the excavation area should be separated from other layers for examination at a later time. For example, when the debris contains materials such as door or window locks, window glass, and interior finish materials, the fire investigator should set these items aside for additional analysis and reconstruction.

As the layers are removed, additional patterns generated by the fire during its development may become visible. For example, investigators may find unburned material below a specific layer in the debris. This could indicate the level at which the ignition occurred. Protected areas should also be examined and documented as part of the documentation of the origin.

Large items found in the debris, such as furniture or appliances, should be left intact if possible. If they must be removed to facilitate the examination of the area, they should be placed in an accessible location for additional examination and potential reconstruction.

 **Fire Scene Reconstruction**

The objective of reconstructing a fire scene is to attempt to re-create the scene as it existed before the fire. Reconstruction of the area surrounding the fire origin is an important step in the investigation process. Key objectives in fire scene reconstruction are as follows:

- Re-create the scene with contents in their prefire location.

- Document fire growth patterns.

- Identify and document potential ignition sources.

- Identify and document first material ignited.

- Document and collect evidence.

Fire scene reconstruction involves the relocation of burned materials and contents, structural elements (wiring), and utility services (piping) back to their prefire position after the fire debris has been removed. Reconstruction not only involves the materials and items identified during debris removal but also includes the identification and relocation of contents removed during overhaul or salvage operations. Materials thrown out during the debris removal are difficult to relocate or analyze during reconstruction. This may lead to the misanalysis of ignition scenarios or potential ignition sources. In such instances, the reevaluation of the debris pile may become necessary.

Successful reconstruction is directly dependent upon the thoroughness and close attention to detail during the debris removal. Information from interviews or prefire photographs can also assist the reconstruction process.

Reconstruction allows the investigator to observe fire patterns, including protected area patterns, in relation to items that were in the area of origin before and during the fire. The analysis of fire patterns conducted on contents outside a structure is not considered a true reconstruction.

Diagrams or notes indicating measurements with the exact location from where the materials were discovered during the debris removal helps to establish the original location of the items. The exact prefire location of contents is often determined by the presence of fire patterns and particularly protected area patterns. The items should be placed and oriented in conjunction with these patterns for the reconstruction to be complete. If the orientation, position, or exact location of an item cannot be determined, the item should not be considered as part of the analysis of the fire patterns.

As this process of debris removal and reconstruction begins, the fire investigator should have developed one or more preliminary fire-spread scenarios. As this process continues, the investigator tests those scenarios against the findings at the scene. Should the evidence indicate that a fire-spread scenario is not probable, other scenarios should be considered. Evidence that supports the final scenario should be collected. If no probable scenario can be determined, additional debris removal and analysis in other areas of the scene may become necessary. As the reconstruction process proceeds, all potential sources of ignition in the area should be identified and ruled out or selected as the most probable heat energy source.

After the fire scene reconstruction is complete, the investigator should document the entire scene through sketches and photographs and then collect relevant evidence. Reconstruction is not only important from the standpoint of visualizing fire patterns but is necessary for conducting a complete and thorough analysis of the fire scene.

# Determining the Ignition Source

After documenting the information related to the area of origin, if it is found, the investigator should explore the area adjacent to the origin for possible sources of ignition. The ignition source of a fire must be one that is able to provide sufficient heat energy to be able to cause ignition of the material first ignited (the fuel located at the point of origin). The determination of the ignition source is one of the findings that the fire investigator will use in determination of the fire cause. Depending on what was the source of heat, the fire investigator may find evidence that is recognizable, such as an electric space heater with combustible waste materials stacked against it. In other cases, the actual heat source may not remain at the point of origin, or it may have been altered or completely destroyed by the fire.

## Ignition Sources and the Fuels They Ignite

NFPA 921 defines a *competent ignition source* as one that has sufficient temperature and energy and is in contact with the fuel long enough to raise it to its ignition temperature. The fire investigator should understand that the physical state of the material first ignited by an ignition source has a significant

impact on the amount of energy required for ignition. For solid fuels, the ignition source must have the potential to supply enough energy over a period of time to cause pyrolysis of the fuel and be at a temperature that is high enough to ignite the resulting gases. Liquid fuels require an ignition source that provides sufficient energy to cause the liquid to change state from liquid to vapor at a temperature high enough to cause ignition of the vapors. Gaseous fuels require the least amount of energy for ignition as there is no energy required to change the fuel's physical state before ignition.

A competent source of ignition for gaseous fuels might be a small spark caused by static or friction that provides enough heat energy to begin the combustion process. The same spark would not be considered to be a competent ignition source for a dense solid fuel, such as a piece of upholstered furniture, a wood stud, or a wooden beam in a building. Gaseous vapors will only be ignited if the ignition source is located so that the concentration of vapor adjacent to it is within the flammable range of the substance. If the vapor concentration is above or below the flammable range, the ignition source will not cause ignition.

Examples of fires where the ignition sources may not be located during debris removal could include:

- Those set with a match or lighter that is then removed from the scene

- Those where the ignition source is consumed by the fire

- In cases involving the ignition of flammable gases or ignitable liquids, the vapors from the fuels may travel a significant distance before an ignition source is encountered by a concentration of vapor within the explosive range. Circumstances such as these will make the job of correlating the ignition source with a material first ignited difficult. In these cases, a careful examination of surfaces in the area of origin may yield fire patterns that could assist in the determination of the cause. A finding that could result from the careful examination of the point of origin is that there is no apparent source of ignition. To make this finding, the fire investigator must rule out any potential sources of ignition at or near the point of origin. A potential ignition source could be ruled out if the fire investigator is able to find evidence that it could not have started the fire. When attempting to rule out a potential source of ignition, the fire investigator may decide that evidence that supports the finding should be documented, collected, and preserved. If specific evidence is not collected, then the available information or data should be documented prior to possible destruction during the removal of debris.

If the fire investigator suspects that the ignition of the fire involved ignitable liquids, samples should be taken for laboratory analysis. Information regarding the collection of samples for laboratory analysis was previously discussed in Chapter 10, "Evidence Collection and Preservation."

As the examination to determine the ignition source continues, the fire investigator should also consider the method of heat transfer for any potential sources of ignition. The investigator must ask how the heat energy transferred from the potential energy source to the fuel. Again, using the knowledge of fire behavior, the fire investigator knows that the methods for the transfer of heat energy are conduction, convection, or radiation.

- *Conduction.* Point-to-point transfer; when the fuel and potential ignition source touch. For example, a portable electric heater with combustible materials pushed against it, or a hot metal surface touching a wooden structural component.

- *Convection.* Heat transferred by gases rising from the fire. An example would be a candle igniting draperies located above the flame.

- *Radiation.* Heat energy transferred by infrared rays from a fire to an opaque object. Probably the best example of this form of ignition is an "exposure" fire. An *exposure fire* is ignited by the radiation from another fire that is large enough and close enough to cause available fuels to generate ignitable gases and then heat those gases to their ignition temperature. Large building fires have been known to generate enough energy to ignite combustible surfaces of other buildings across streets or other large open spaces.

## Evaluation of Potential Ignition Sources

As debris is removed and the area or point of origin is exposed, the fire investigator must begin a careful examination of the surrounding space for potential sources of heat energy. As previously discussed, the source of the heat of ignition is an energy source that is capable of providing the heat energy necessary to ignite the material first involved in ignition. NFPA 901, *Standard Classifications for Incident Reporting and Fire Protection Data*, lists the following potential forms of heat of ignition:

Heat, sparks, ember, or flames from outside, open fires

Heat from fuel-fired or fuel-powered equipment (gas or liquid fuel)

Heat from fuel-fired or fuel-powered equipment (solid fuel)

Heat from electrical equipment

Heat from hot object

Heat from explosive or fireworks

Heat from other open flame, sparks, or smoking materials

Heat from natural source

Heat spreading from separate fire source

### Heat, Sparks, Ember, or Flames from Outside, Open Fires

Open fires outside a building can generate sparks or embers that could result in the ignition of combustible materials used as structural components. One example would be sparks or embers from an outside fire igniting wood shingles on the roof of a structure. A flaming wood ember could reach temperatures of approximately 1880°F (1 027°C). While this temperature is high enough to ignite a wood shingle or other combustible material, the fire investigator should also evaluate the distance the ember had to travel before landing on the combustible surface, as the ember loses energy as it travels. If the outside fire is close to the point of origin — within 30 to 40 feet (9 to 12 m) — the possibility of a burning ember serving as a competent ignition source exists. In this case, the type of heat transfer would be convection, while the method of ignition would be conduction.

Outside fires can also result in the ignition of combustible surfaces as a result of heat energy radiated from the fire to the target fuel. Again, the fire investigator must consider the distance from the original fire to the target fuel when evaluating this potential ignition source. As the distance from the source to the target increases, the potential for ignition decreases. The heat release rate of the source (how big the fire is) and the distance to the target fuel will have to be considered carefully to identify this type of ignition source.

### Heat from Fuel-Fired or Fuel-Powered Equipment (Gas or Liquid Fuel)

Ignition sources in this category include cutting or welding operations using gas-fueled torches, plumber torches, and heat or sparks from gas- or liquid-fueled equipment (including pilot lights). Welding and cutting operations and the joining or separating of metals using a gas-fueled torch are well recognized as significant hazards of ignition. These operations are a hazard due to the fact that at least two sides of the fire tetrahedron — heat and air — are always present during the process. If there are combustibles near the operation, there is a very good possibility that an ignition could occur. This source of heat energy is one of those not usually found during the debris removal and point of origin examination process. The determination will almost always be made based on the evidence at the point of origin and knowledge that hot work was conducted in the area before the ignition of the fire.

The flame temperatures of an oxygen/gas mixture are always above the ignition temperature of common combustibles. Thus, the flame from the device is a potential source of ignition should it come into contact with a fuel source. During cutting operations, the hot molten metal resulting from the process generates thousands of potential ignition sources. Sparks and molten metal can travel many feet (meters) from the work area and cause ignition. An ignition from the torch flame would be immediately adjacent to the work area. As a reference, NFPA 51B, *Standard for Fire Prevention During Welding, Cutting, and Other Hot Work*, indicates hot work procedures for cutting and welding require that an area of 35 feet (11m) around the operation be free of combustibles (or combustibles in the area covered with flame-resistant materials or guards) and that a fire watch be posted for at least 30 minutes after the operation is completed. The same potential exists where plumber torches are used to solder pipes. If the area is not protected, the flame may ignite combustible materials adjacent to the pipe. The potential for drops of solder to ignite combustibles is low as the melting temperature of the material is in the 275°F to 350°F (135°C to 177°C) range which is at or below the ignition temperature of wood and paper (Table 14.1).

### Table 14.1
### Time Required to Ignite Wood Specimens

| Wood 1¼ in. × 1¼ in. × 4 in. (32 mm × 32 mm × 102 mm) | No Ignition in 40 Min | | Exposure Before Ignition, by Pilot Flame, Minutes | | | | | | |
|---|---|---|---|---|---|---|---|---|---|
| | °F | °C | 356°F (180°C) | 392°F (200°C) | 437°F (225°C) | 482°F (250°C) | 572°F (300°C) | 662°F (350°C) | 752°F (400°C) |
| Long leaf pine | 315 | 157 | 14.3 | 11.8 | 8.7 | 6.0 | 2.3 | 1.4 | 0.5 |
| Red oak | 315 | 157 | 20.0 | 13.3 | 8.1 | 4.7 | 1.6 | 1.2 | 0.5 |
| Tamarack | 334 | 167 | 29.9 | 14.5 | 9.0 | 6.0 | 2.3 | 0.8 | 0.5 |
| Western larch | 315 | 157 | 30.8 | 25.0 | 17.0 | 9.5 | 3.5 | 1.5 | 0.5 |
| Noble fir | 369 | 187 | — | — | 15.8 | 9.3 | 2.3 | 1.2 | 0.3 |
| Eastern hemlock | 356 | 180 | — | 13.3 | 7.2 | 4.0 | 2.2 | 1.2 | 0.3 |
| Redwood | 315 | 157 | 28.5 | 18.5 | 10.4 | 6.0 | 1.9 | 0.8 | 0.3 |
| Sitka spruce | 315 | 157 | 40.0 | 19.6 | 8.3 | 5.3 | 2.1 | 1.0 | 0.3 |
| Basswood | 334 | 167 | — | 14.5 | 9.6 | 6.0 | 1.6 | 1.2 | 0.3 |

Heat from pilot lights and fuel-fired equipment is also a common source of ignition for fires. To make a determination that a fire was ignited from this type of source, the fire investigator should consider the location of the fuel first ignited and the method of heat transfer from the source to the fuel (Figure 14.1). If the fire investigator suspects that the source of heat was a fuel-fired device, care should be taken during the debris removal process to protect the remains of the device and the area adjacent to it. As the examination proceeds, the fire investigator will have to determine whether additional expertise is required to examine the unit before disturbing it at the scene. Whatever decision is made, the equipment should be photographed and documented before its removal. If the unit is a suspected ignition source, care should be taken to prevent spoliation throughout the process of debris removal and evidence collection, including avoiding the movement of knobs, switches, and valves.

**Figure 14.1** Clothes within gas-fired dryers are often the material first ignited. *Courtesy of Russ Chandler.*

## Heat from Fuel-Fired or Fuel-Powered Equipment (Solid Fuel)

These sources of heat energy include wood and coal heating equipment, the chimneys servicing those devices, and sparks or embers escaping from the device or chimney. Because these devices are designed to generate heat using solid fuel, they are potential sources of ignition if combustibles come into contact with the hot surface of the device or its chimney. The method of heat transfer in these cases is conduction. The improper installation of a device or its chimney that places a hot surface too close to combustible materials may also be a potential source of ignition through radiant heat transfer. Another potential source of heat energy could be burning or glowing pieces of the solid fuel falling from the unit onto nearby combustibles.

If a heat-producing device is located at or near the area of fire origin, the investigator should examine the device and the associated piping and chimney as a potential source of ignition. The device should be well documented either as a source of ignition or to provide evidence to rule it out later in the investigation.

## Heat from Electrical Equipment

Electrical equipment that fails is a potential source of heat energy sufficient to start a fire. Given the reliance of our society on electricity, finding electrical wiring and or electrically operated devices at or near the point of origin is to be expected. The fire investigator performing the scene examination has the task of evaluating these potential heat sources and determining whether they are potential sources of ignition or can be eliminated (Figure 14.2).

**Figure 14.2** Electrical appliances may be a source of ignition.

The initial step in the examination of electrical wiring or equipment located near the point of origin is to determine whether there was a failure, a possible ignition source of the fire, or just the result of fire temperatures. Examples of physical evidence of electrical failures include:

- Copper conductors that have melted and resolidified in a spherical shape

- Switch contacts that are fused together

- Electrical insulation surface adjacent to a conductor that has been hotter than on the surface exposed to the fire

- Any metal deterioration not believed to have resulted from oxidation due to the fire and overcurrent devices (fuses or circuit breakers) that have operated

The investigator is cautioned that every one of the previous examples could be the result of a fire and not the ignition source.

The following are some common examples of how electricity can cause an ignition source.

- **Short circuits.** A short circuit is an abnormal path of current in a circuit that normally leads to an overcurrent condition. If the short circuit is the result of contact between metal objects normally insulated from one another, the current could increase to the point that metal is melted. If the metal is heated to the point that it is vaporized, an arc occurs. While the arcing is very brief, droplets of molten metal (sparks) will be thrown from the site. Depending on the fuels at or near the point of the arcing, an ignition could occur. Where the failure involves a copper conductor, the metal droplets will cool very rapidly as they travel through the air. Molten droplets from aluminum conductors burn as they travel through the air and have a greater ability to cause the ignition of combustible materials on which they land.

The ability of metal droplets to cause ignition depends on the available fuel and the size and heat content of the molten particles. Most droplets will only be capable of igniting fuels with a high surface-to-mass ratio (fine fuels — such as dust or thin paper) or those already heated almost to the point of ignition.

- **Poor connections.** The most common cause of electrical fires after user abuse of electrical equipment is poor electrical connections. The investigator should examine all connections near the determined area or point of origin. Partially melted screw heads, copper conductors melted under a wire nut, and eroded steel lug screws are not normal results of a fire environment. Shorting and/or arcing in a branch circuit is to be expected — a localized overtemperature condition at a receptacle connection is not normal. Examining connections involving aluminum conductors requires special care. If the surface of an aluminum conductor is exposed to air, a high-resistance oxide forms. Current flow through the oxide layer causes localized but very high temperatures. This spot heating could be enough to result in ignition of combustibles that are very close to the connection point. Of course, aluminum melts even at temperatures generated in a typical residence fire. Therefore, the investigator will have to rely on damage patterns to screw heads, lug screws, or other metal objects in evaluating the prefire connection quality.

- **Resistance heating.** The flow of current through a conductor normally produces some heat due to the resistance of the material used to form the conductor. This heat is normally dissipated to the surrounding air. If an excessive amount of current flows through the conductor, enough heat may be liberated to ignite adjacent combustibles or damage electrical components or the wiring in appliances.

The previous list is aimed at abnormal conditions that result in heat being generated by the electrical energy. Normal heating from stoves, heaters, lightbulbs, etc., is expected and would not be considered an electrical source of ignition. During this analysis, the fire investigator should also remember that many devices are designed to have current levels that could result in enough energy to result in ignition if they are used improperly. A 100-foot (30 m) heavy-duty extension cord may be capable of safely providing electrical energy to a window air conditioner as long as it is fully extended. The same cord tightly coiled on the floor near an upholstered chair could generate sufficient energy to both pyrolyze the fabric and damage the extension cord and insulation. An electric arc could then result in the ignition of the chair.

The investigator should also be aware that different metals may combine at fire temperatures and form alloys that have characteristics different from any of the original materials. While copper melts at 1981°F, a temperature not generally reached throughout a typical structural fire, it will melt at much lower temperatures if contacted by molten aluminum. Zinc and other so called "pot metals" can have the same effect. The aluminum and heated copper alloy at the copper surface form a material that has a lower melting point than either aluminum or copper. Severed branch circuit conductors that have a silvery appearance most likely have been subjected to this "eutectic" alloying, not electrical arcing.

When evaluating a potential electrical source of ignition, the fire investigator must determine whether the heat generated was sufficient to cause ignition of the materials first ignited. If the material first ignited is solid, a competent electrical ignition source must be capable of developing ignitable vapors by pyrolysis and then become hot enough to ignite those vapors.

## Heat from Hot Object

A hot object is one of the most common sources of heat resulting in ignition. Examples of this source of heat energy include:

- Properly operating electrical devices, such as an electric heater or lightbulb, with combustibles touching or very close to the unit

- Hot metal fragments from machinery

- Hot ashes that are improperly discarded

- Heat generated by friction, such as overheated fan belts, hot boxes on railcars, or tires

As with the evaluation of any other potential source of heat energy, the fire investigator has to determine whether the hot object was a competent source of ignition for the fuel that was first ignited. This would include determining the autoignition temperature of the material and evaluating the heat source to determine whether the conditions were present for these tempera-

tures to occur. The method of heat transfer should be determined and also evaluated. If the fuel was in direct contact with the hot object (heat source), the method of transfer was conduction. If there was a distance between the fuel and the heat source, the transfer was most likely by radiation and would require more energy than transfer by conduction.

When evaluating electric lighting as a potential source of ignition, the fire investigator should note the type of bulb involved and how the bulb was oriented to the fuel. A low-wattage incandescent lightbulb can generate surface temperatures of up to approximately 500°F (260°C). The surface temperature of an incandescent bulb depends on its wattage and position (that is, base up, base down, etc.) The normal surface temperature of an incandescent bulb is not typically a competent ignition source (depending on its wattage and position) for most fuels. However, if the bulb is insulated by being wrapped in bedding or in a towel, it could generate temperatures high enough to ignite paper or cloth.

High-intensity lamps that use quartz halogen bulbs are a very competent ignition source because the bulbs used will generate surface temperatures of up to 1650°F (900°C). A quartz halogen bulb in contact with or in close proximity to combustible materials could result in flaming ignition.

## Heat from Explosive or Fireworks

Items manufactured to produce sufficient heat of energy to cause metal to glow or to result in a sudden release of energy provide a viable ignition source, particularly in contact with low mass fuels. Items included in this category include:

- Heat generated by munitions

- Tracer ammunition

- Blasting agents

- Fireworks (Figure 14.3)

- Incendiary devices

- Model rockets

These sources of heat energy are, for the most part, designed to burn or actually cause an ignition. As such, they can be expected to be competent sources of heat energy.

**Figure 14.3** Fireworks are a frequent source of accidental ignition.

The fire investigator examining the point of origin of a fire suspected of being ignited by explosives or fireworks should attempt to locate portions of the device during the debris removal process. If an incendiary device was used, samples at and around the point of origin should be taken for laboratory analysis to determine the type of material used in the device. Explosives or fireworks will also leave traces that can be identified in the laboratory.

Not all incendiary devices are constructed using ignitable liquids as the initial fuel source. Some devices use materials that would normally be found in the location of the fire. One example might be a high-wattage lightbulb with matches attached to it, insulated by being covered with shredded paper or wrapped in cloth. Careful examination of the debris and good documentation of the findings assist the fire investigator as the investigation progresses in making the determination that the fire was intentionally set.

### Heat from Other Open Flame, Sparks, or Smoking Materials

Items requiring an open flame to function or when functioning results in the production of flame, heat or sparks are common ignition sources. Sources of heat energy in this category include:

- Cigarettes
- Cigars or pipes
- Matches
- Cigarette lighters
- Candles
- Warning flares  (Figure 14.4)
- Backfires from internal combustion engines

Smoking materials including cigarettes, cigars, matches, and lighters are commonly found to be ignition sources. As the fire investigator examines the point of origin, a determination of the material first ignited should be made if possible. If smoking materials are suspected as the energy source for the fire, a determination of how the source came into contact with the fuel should be explored. A cigarette ember can generate temperatures from 930°F to 1300°F (500°C to 700°C) while free burning. When air is drawn through the cigarette, temperatures generated could be as high as 1670°F (910°C). A lighted cigarette placed on a surface may cause only scorching of that surface if it is uninsulated. The same scenario with the cigarette insulated with even a single layer of cloth, such as a bed sheet, could generate sufficient heat to result in ignition. Cigarette ignitions of upholstered furniture almost always require that the cigarette is located so that it is partially between the cushions with a sufficient air supply to allow it to burn while the heat generated is absorbed by the cushion and padding. These ignitions require a significant period of time before flaming ignition occurs. Laboratory burn tests show that times from two minutes to several hours are necessary for flaming ignition to occur. The time required is dependent on the position of the cigarette and the materials used to construct the furniture item. The fire investigator should also note that during the pre-ignition process, a significant amount of smoke can be generated and result in harm to occupants of a structure who are not aware of the situation. When attempting to identify or rule

**Figure 14.4** Flares are often used to intentionally ignite combustibles.

Chapter 14 • Debris Examination, Removal, Reconstruction, and Determination of Ignition Source **185**

out smoking materials as the ignition source, the fire investigator will have to carefully examine and document the point of ignition.

Candles are also a common source of heat energy. They produce sufficient energy to cause ignition of nearby combustible materials and by design can produce that energy for a substantial period of time. Methods of heat transfer can be where fuel is in contact with the candle (conduction) or where the heat from the burning candle comes into contact with combustibles above the flame (convection). For a candle flame to result in an ignition of materials from radiation, the fuel would have to be in very close proximity to the flame. Where candles are involved in the ignition, the fire investigator may find wax remnants at or near the point of origin. The examination of this area should also provide an indication of the material first ignited as well as burn patterns that point to the candle as the source of heat energy.

## Heat from Natural Sources

Ignition scenarios, although relatively rare, occur in nature and require no human intervention. Sources of heat energy in this category include:

- Heat from the sun
- Spontaneous heating
- Static discharge
- Lightning strike (Figure 14.5)

Under normal conditions, the heat energy received from the sun is not sufficient to cause ignition. Where this energy is focused onto a small area, such as a magnifying glass aimed at a piece of paper, sufficient energy is available to cause pyrolysis and potentially ignition. Examples of this type of fire would be where a magnifying mirror is arranged so that the sun's energy is focused on a combustible surface, such as draperies, wood interior finish, or paper, for a period of time. During the examination of the point of origin, the item that caused the focusing of the energy should be identifiable. Likewise, the availability of sunlight to the item should be well documented.

*Spontaneous heating* is a process through which an organic material increases in temperature without an external source of heat energy. Spontaneous

heating is the result of the oxidization of a material and the generation of heat as a result of that process. Table 14.2 provides examples of materials prone to self-heating. Other materials are listed in references such as the *Fire Protection Handbook*.

While self-heating is most common in organic materials such as those listed in the table, some metal shavings or powders are known to also generate heat as a result of oxidization. In isolated instances, these materials may generate sufficient temperatures to self-ignite.

**Figure 14.5** Lightning is one form of natural ignition.

| Table 14.2 Examples of Materials Subject to Self-Heating | |
|---|---|
| **Material** | **Potential for Self-Heating** |
| Charcoal | High |
| Fish meal | High |
| Linseed oil rags | High |
| Foam rubber | Moderate |
| Hay | Moderate |
| Manure | Moderate |
| Bailed rags | Moderate to Low |
| Sawdust | Low |
| Grain | Low |

Source: *Fire Protection Handbook*, 18th Edition, 1997, p. A-15)

Factors that influence the ignition of combustibles by self-heating include:

- **The rate of heat generation.** The heat generated by the material must be at a rate that is greater than the rate heat is being dissipated to the surroundings. The higher the temperature of the material, the faster heat will be generated by self-heating. Thus, if the material was warm or hot when stored, the higher the potential for self-heating and ignition.

- **Effects of ventilation.** For ignition to occur as a result of self-heating, there must be enough air available to support the oxidization process but not too much as to allow all the heat to be dissipated by convection. An excellent example of this is a linseed oil soaked rag. Crumpled up and placed in a container, it has a great potential to reach the point of self-ignition. The same rag spread out and placed over the edge of the container so that any heat generated is allowed to dissipate with the help of air movement will not normally generate sufficient heat to self-ignite.

- **The insulation properties of the immediate surroundings.** For self-ignition to occur, the material must be insulated so that the heat generated by the oxidation process does not dissipate. The previous crumpled rag example results in insulation so that the heat generated inside of the ball is not easily dissipated. Piles of materials also provide insulation so that self-heating and ignition could occur deep within the pile if sufficient air to support it is available.

During the examination of a scene where spontaneous heating is a potential source of ignition, the investigator must identify the suspected agent or material involved. He must then determine its potential for self-heating while attempting to document the conditions that could have allowed the ignition to occur.

Lightning is a natural static discharge that is capable of delivering very high levels of electrical energy to localized areas. Lightning commonly strikes the tallest object in its path to the ground and can enter a structure through several different paths including:

- Striking metal objects on a structure — antennas, masts, rooftop HVAC units, or other components of the structure that extend higher than the roof

- Direct strikes to the structure

- Striking tall objects (trees, other structures, etc.) that are near the building and traveling horizontally into the building

- Striking power, telephone, or cable lines and connections traveling into the structure via the service connection

As the lightning bolt travels along a conductive path, it may separate as additional conductive paths are encountered. For example, the lightning bolt may travel from electrical wiring to the building's plumbing system if the pipes are used as a ground for the electrical system.

Lightning delivers a tremendous amount of energy to the location of a strike. This may result in damage to conductors because of a high overcurrent condition resulting in severe damage, which includes melting, breaking, or vaporization of the conductors at several locations along the path of travel. The energy delivered by the strike may also result in explosive destruction of objects or structural components in the path of travel.

*Static electricity* is a term used to describe electricity that is trapped or prevented from escaping from the surface of an object. A body or object with an accumulation of trapped electrical energy is said to be "charged." That is, the object will be a positive or negative differential as compared to its surroundings and therefore subject to discharging its potential energy if a path to ground is provided.

Static electricity is normally generated by the motion of two dissimilar bodies in contact with each other. These bodies can be either solid or liquid, and one or both will normally be a poor electrical conductor. As the bodies move against each other, electrons are stripped from one body and collect on the other. This action causes a positive charge on the body losing the electrons and a negative charge on the body that collects the electrons. Examples of actions that could generate static electricity include:

- Motion that involves friction between two dissimilar surfaces that are in direct contact.

- Gases that contain particulate matter flowing from an opening in a pipe or hose. This could be water droplets in steam or solid particles in a gas, including air.

- The point of separation of a stream of liquid from a spout, nozzle, faucet, or hose.

- Pulverized materials passing through a chute or pneumatic conveyor.

- Nonconductive belts that are in motion. Including fan belts, conveyor belts, or drive belts used to power machinery.

For ignition to occur as a result of a static discharge, the following conditions must be fulfilled:

- Static electricity must be generated.

- Separate static charges must be accumulated, and the electrical potential maintained.

- There must be a discharge arc of sufficient energy to cause ignition of the fuel.

- There must be an ignitable mixture of fuel in close proximity to the discharge arc.

When investigating fires where a static discharge is suspected of being the ignition source, the fire investigator involved in the examination of the point of origin would have to develop evidence based on the previously listed conditions to support the finding. Some physical evidence may be available to identify a source of static electricity, such as a machine that is driven by a fan belt or evidence of a liquid being poured from a container. Other evidence that supports the finding may be circumstantial and based on interviews and other data collected after the detailed scene examination. At the scene, the fire investigator should document any available information regarding the source of the discharge and the fuel first ignited.

For a static discharge to result in ignition, the resulting spark must have sufficient energy to cause the ignition of fuel that is in close proximity to the discharge. The types of fuels that are most susceptible to ignition from static would be ignitable liquid mists or vapors, ignitable gases, and combustible dusts. From the study of fire behavior, the reader should remember that fuels with a high surface-to-mass ratio require less energy for igni-

tion. Thus, static sources would not be expected to be competent sources of heat energy for very dense solid fuels. However, they are potentially competent heat sources where solid fuels are finely divided, such as wood or other combustible dusts. Fine mists of ignitable liquids may also be ignited by static sources as well as gases or vapors that are within their flammable limits.

### Heat Spreading from Separate Fire Sources

Fires often ignite secondary fuel packages, and the investigator must determine whether apparent multiple fires have separate origins or are a result of natural fire propagation. Sources of heat energy in this category include:

- Direct flame impingement from another fire

- Radiated heat from another fire

- Heat from flying brands, embers, or sparks generated by another fire

- Conducted heat from another fire

Heat from fire exposure is a viable source of heat energy for the ignition of additional fires at some distance from the original fire. When this source of ignition is suspected, the source of energy should be apparent, and the fire investigator should document the method and path of heat transfer to the point of fire origin of the fire being investigated.

 **Conclusion**

At the completion of the debris removal and reconstruction phase of the scene examination, the fire investigator should be able to develop a probable scenario regarding the ignition of the fire (ignition source and material first ignited) and the point(s) or area(s) of origin. While additional analysis of the available information and evidence may be necessary before a final determination of cause can be made, the fire investigator should have sufficient information — photos, diagrams, field notes, and physical evidence — from the fire scene to move into the analysis phase of the fire investigation.

# The Analysis of Investigative Findings

Up to this point in the investigation of a fire incident, the fire investigator has been primarily focused on the fire scene with activities that involve scene documentation, the collection of physical evidence, and taking statements from witnesses. Once the thorough examination of a fire scene has been completed, the investigator moves into a phase of the investigation that involves the organization and analysis of the available information with the ultimate objective of determining the origin and cause of the fire. When it is determined that a fire was intentionally set, the fire investigator may also be involved in determining who was responsible for the fire.

The analysis phase of the fire investigation can be very simple and may be completed during the scene examination if the origin and cause are easily determined. When the determination of the origin and cause is not readily made during the examination of the scene, the post-scene investigation can be complex and time-consuming. The post-scene investigation may include compilation of records such as:

- Police reports regarding the incident

- Fire department incident reports

- Reports developed by private investigators

- Photos and videos obtained from investigative and other sources, such as television stations or bystanders at the fire scene

Depending on the incident complexity, other elements of the post-scene investigation may include:

- Additional interviews of involved parties or individuals who might possess information regarding the investigation

- Submitting evidence for laboratory analysis and relating the results to the fire scene

- Developing data necessary to assist in the determination of the fire cause

- Seeking experts to assist in the development and/or analysis of investigative information

- Review and analysis of records and documents related to the incident and/or the involved parties

This chapter provides the fire investigator with the basic information required to organize and plan a post-scene investigation and analyze the information gathered as part of the fire investigation.

## Analysis of Investigative Findings

The objective of the post-scene investigation is to develop additional information regarding the origin and cause of a fire. The need for this information is determined after a careful evaluation of the information obtained from the scene investigation. To make a final determination of the origin and cause of a fire, the fire investigator needs to answer each of the following questions:

- What factors or circumstances brought the components of the fire tetrahedron (heat, fuel, and oxygen) together?

- What was the spread of the fire once it was ignited? How did the fire spread? Were the fire protection systems adequate? Did the building construction and interior finish or contents contribute to the spread of the fire?

- What factors were responsible for injuries or fatalities resulting from the fire? These factors may include the adequacy of the means of egress, fire alarm systems or defend in-place building components, and the role of specific products that may have emitted toxic by-products during the fire. If there were firefighter injuries or fatalities, the factors that contributed to those injuries should also be determined.

- What human factors were involved in the ignition of this fire? These factors may include fires that result from unsafe acts or negligence as well as those that are maliciously set.

If it is determined that the fire was intentionally set, the fire investigator, in most jurisdictions, will then be involved in determining the party or parties who were responsible for the fire.

## Cause Classification

The primary objective of the fire investigation is the determination of the cause of the fire or explosion. *Fire cause* can be defined as the sequence of events that allow the source of ignition and the fuel to come together. NFPA 921 lists the following potential classifications of fire cause:

- **Accidental**. Those fires that do not involve a deliberate human act to ignite or spread the fire into an area where the fire should not be.

- **Natural**. Fires, such as those caused by lightning, storms, or floods, where human intervention has not been involved in the ignition process.

- **Incendiary**. A fire deliberately set under circumstances in which the responsible party knows it should not be ignited.

- **Undetermined**. The classification used when the specific cause of a fire cannot be determined. This classification may be used as an interim classification as the fire investigation is proceeding, or it may be the final outcome if additional information cannot be developed that identifies the specific cause. This does not, however, mean that if each of the specific components of the fire tetrahedron is not specifically identified, the fire should be called undetermined. A cause should be given if sufficient information is available. For example, when there is a known natural gas leak that resulted in an explosion and the specific source of ignition is not known, the fire investigator should assign the fire to the accidental classification and list the source of ignition as undetermined. Also, when the evidence points to the intentional use of ignitable liquids as an accelerant and the specific ignition source is not determined or found, the fire investigator should assign the fire to the incendiary classification and list the source of ignition as undetermined.

In the past, fire investigators often used the term *suspicious* to describe fires when the specific cause could not be determined but was thought to be intentional. Investigators should avoid the use of this term. When there is not enough information available to make a specific determination, the investigator should classify the fire as undetermined. If it only *appears* that the fire was incendiary in nature, additional information or evidence must be sought during the post-scene investigation to support an incendiary finding.

## The Collection and Organization of Investigative Information

The process of collecting investigative information begins when the investigator is first assigned to an incident. The information the investigator collects may be in the form of notes, reports, sketches, evidence packages, film, videotape or audiotape of interviews, and more. The first task in the post-scene investigation phase is to assemble all this information into a format that can be evaluated and used during the investigative process.

The organization of investigative data may be as simple as identifying each specific component of the file as it exists and inserting each into file folders or a three-ring binder with dividers that identify the individual items. A complex case may require the use of computers to organize and store information (Figure 15.1). Whatever method is used, the objective is to organize the information that is currently available into a format so that it is accessible to the investigators working on the case. All evidence that has been collected must be properly documented and secured so that it is available for use in the investigation and subsequent court proceedings. As discussed in Chapter 10, "Evidence Collection and Preservation," any evidence collected must be properly secured and the chain of custody maintained.

Once all the information collected is organized, the investigator should review and evaluate the material. At this point, the investigator determines exactly what information is available in the file. This information identifies information that is necessary to complete the investigation and can be used in the planning process for the post-scene investigation. The investigator may use printed forms to compile the collected information, depending on the rules of the jurisdiction in which he is working. If a form is not used, the investigator

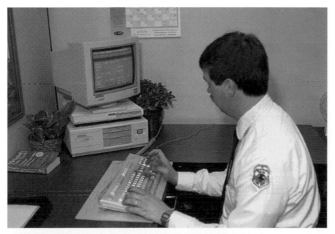

**Figure 15.1** Some cases may require the use of a computer to track data.

may make a simple list of the contents of the investigative file to document what is available and identify information or reports that must be obtained for further evaluation.

 ## Planning the Post-Scene Investigation

With all the available investigative information organized and cataloged, the investigator can then begin to develop a plan for the continued post-scene investigation. Based on the evaluation of the investigative file, the fire investigator should have an understanding of the information that is currently available and of the potential scenarios that are being evaluated. The investigator should identify and obtain for the file any information from other sources, such as fire and police department reports or insurance reports, which could assist the investigation. The investigator should also determine what information is needed to prove or rule out the ignition scenarios that are being evaluated. This portion of the plan may identify any further laboratory work that will be necessary for the analysis of evidence collected at the fire scene as well as special expertise needed to support the investigation.

 ## Forensic Laboratories

The forensic laboratory is a tool used to assist in the determination of the cause of a fire. The types of analysis that are typically requested by the fire investigator include:

- **Fire debris analysis** — Determines the presence of materials that may be found at the fire scene, including traces of ignitable liquids that were used as an accelerant

- **Microanalysis** — Examines items such as damaged electrical wiring, tool marks, and impressions from tires and shoes found at a fire scene; also involves the analysis of broken glass, smoking materials and matches, and hair or fibers found at the scene or on a suspect

- **Latent fingerprint identification** — Compares fingerprints found at a fire scene (at points of entry, on fuel containers, on a vehicle, or on incendiary devices) to those of a suspect

- **Serology**—Analysis of body fluids such as saliva, blood, urine, or semen

- **Forensic anthropology** — Used for the identification of deceased victims

- **Forensic pathology**—Determines the cause and manner of death

- **Firearms analysis** — Examines firearms or projectiles for classification and/or identification

Once the fire investigator determines that materials require laboratory analysis, he may have a choice in laboratories. The following criteria may assist in the selection process:

- The laboratory and scientist have a reputation of excellence and are certified or accredited by the appropriate agencies.

- The laboratory employs state-of-the-art techniques.

- The scientist understands and uses proper techniques for handling evidence.

The reliability of the results should be the primary concern when selecting a laboratory to perform forensic analysis of evidence collected as part of the fire investigation. The forensic science community has developed standards and certifications that can assist the fire investigator in the evaluation process. Examples include:

- Association of Crime Laboratory Directors (ASCLD), which provides criteria used to judge whether a facility and its practices provide an atmosphere conducive to the quality of work necessary in the forensic field

- Accreditation Board of Engineering and Technology (ABET)

- The American Society of Testing and Materials (ASTM), which maintains a committee on Forensic Sciences (E-30) that establishes standard methods, practices, and guidelines for the forensic field

- American Board of Criminalistics (ABC), which is a national peer review group that certifies forensic scientists in specific disciplines including fire debris analysis

Laboratories that follow national standards, use recognized procedures, and employ certified fo-

| Table 15.1 Common Laboratory Tests Used to Support Fire Investigations (Partial) |
|---|
| **Test Method** |
| Gas Chromatography (CG) |
| Mass Spectrometry (MS) |
| Infrared Spectrophotometer (IR) |
| Atomic Absorption (AA) |
| X-Ray Fluorescence |
| Flash Point by Tag Closed Tester (ASTM D 56) |
| Fire and Flash Point by Cleveland Open Cup (ASTM D 92) |
| Flash Point by Pensky-Martens Closed Tester (ASTM D 93) |
| Flash and Fire Point of Liquids by Tag Open Cup Apparatus (ASTM D 1310) |
| Flash Point by Setaflash Closed Tester (ASTM D 3828) |
| Autoignition Temperature of Liquid Chemicals (ASTM E 659) |

*Source: NFPA 921, Sec. 9-10.*

rensic scientists demonstrate their commitment to providing high-quality results. If possible, before selecting a forensic laboratory, the fire investigator should check the references provided by the laboratory as well as with any peers who may have had dealings with the facility in the past.

Table 15.1 lists some but not all of the tests that can be provided by the forensic laboratory.

##  Special Experts

The investigation of fires may involve the development and analysis of complex information that is beyond the expertise of the fire investigator assigned to an incident. Examples that might require support from special experts include:

- Examination and analysis of electrical equipment or components

- Examination and analysis of building systems or components

- Examination and analysis of fire protection suppression and detection systems

- Examination and analysis of industrial or manufacturing processes
- Examination and analysis of financial records and data

The qualifications of experts required to support the fire investigation depend on the type of assistance needed. This required expertise could include:

- Engineers with a background in one of the major disciplines such as electrical, mechanical, chemical, civil, or fire protection
- Industry experts for assistance in incidents that involve a specialized industry, process, or piece of equipment and who may be associated with trade associations or work for companies that produce machinery and processing equipment
- Forensic accountants to assist in the acquisition and analysis of financial data
- Attorneys for guidance in specialized legal areas
- Insurance representatives to research information, including coverage and prior losses and claims
- Polygraph operators to administer and interpret polygraph tests
- Physicians and forensic pathologists
- Psychologists or behavioral disorder specialists

Information obtained during the post-scene investigation is added to the investigative file for use in the analysis of the incident and the ultimate determination of cause. The accuracy of the fire scenario must be continually reevaluated, and the investigator must be willing to change the scenario based on new or conflicting information.

 ## Analytical Methods Used in Fire Investigation

The ability to analyze investigative information is a critical skill required of the investigator. Analysis of information may be a simple task for most fire investigations when the volume of information is relatively small. As the complexity of an investigation increases, so does the task of sorting through the information and making decisions regarding its relevance. The data must be organized to be manageable, and a formal analysis process must be followed for the results to be meaningful. A systematic approach allows the fire investigator to obtain a better overview of the available data, identify relevant information, and use that information to make informed decisions related to the investigation. The use of analytical investigative techniques may increase the quality of the fire investigator's work product while reducing the time required to perform essential tasks.

## Graphical Representation

Many of the analytical tools that the fire investigator uses involve the graphical representation of incident-related data. The importance of sketches and drawings has already been discussed. There are many other areas where diagrams, graphs, or charts may be used to graphically represent large amounts or complex data in a relatively small space and may make the information easier to understand. Graphs are excellent tools for the representation of numerical data. Examples of graphs the fire investigator might use would be time versus temperature for a fire, water flow over time for an automatic sprinkler system, profit and loss in dollars over time for a business. High-quality graphs may be developed by hand using graph paper and drawing tools or generated using any of a number of easy-to-use software programs designed for personal computers. Graphing may also be useful in the analysis and presentation of the information to other investigators, supervisors, or a jury. While not all information lends itself to graphical representation, the fire investigator should look for methods to organize the data in a manner that is easy to understand and use in the ongoing decision-making process.

### Time Lines

Time lines are a special form of graph used to depict the occurrence of events over a period of time, such as seconds and minutes. Once a time scale is selected, a decision must be made on the reference times that will be used. For example, if the investigator is attempting to develop a time line from fire ignition to the time the fire was extinguished, there may be a number of times available from different sources. Times recorded by each source of information may not be in sync with each other. In that case, the source with the most information, most likely the fire department dispatch center, might be selected as the reference time.

A well-designed time line with carefully correlated times and events is a valuable tool for the analysis of fire and explosion incidents. It provides a very understandable visual representation of known events and potential scenarios that lead up to those events. Time lines of other types of events with longer time periods, such as building ownership and changes in insurance or financial status of individuals or businesses, can also be very useful in the analysis phase of the investigation.

## Link Analysis

As described in the *BATF Fraud Investigation* manual, *link analysis* is a method of computing, organizing, and utilizing data relating to an investigation. It allows the analysis and presentation of complex data in a clear and concise manner. Link analysis is a graphical representation of relationships between two or more persons or organizations and may involve relationships through telephone numbers, street addresses, or vehicle license plates, etc.

## Event Flowcharting

*Event flowcharting* chronologically displays the movements of events or occurrences either over time or through a system. An example of event flowcharting is the event displaying of gambling debt, followed by the loss of job, marriage breakup, and fire.

## Commodity Flowcharting

*Commodity flowcharting* portrays the movement of a tangible item, such as money or stolen property, through a system. A good example of commodity flowcharting would be the portrayal of the movement of items removed from a structure before a fire and placed in storage, pawned, or sold at a flea market.

## Financial Profiling

*Financial profiling* is an investigative tool that allows the investigator to organize the financial data of an individual or organization into a graph or chart. This allows the information to be displayed, and any deviations that may establish a pattern of financial behavior become easier to detect.

## Inference Development

This is the development of a meaningful inference based on the supporting premises using one or more of the techniques previously discussed. The inference may be staged in the form of a hypothesis, conclusion, prediction, or estimate. The inference will be based on the data available and the knowledge, experience, and expertise of the investigator.

This is a subjective process, and the resulting value is dependent on the level of confidence the investigator has in the inference. The inference gives direction to the investigation and should be used to test the validity of the information and provides for recognition of missing data.

##  The Incendiary Fire

Incendiary fires are those fires that are maliciously and deliberately ignited. Descriptions of some of the key indicators of incendiary fires follow:

- **Multiple points of origin** — One must be able to prove that the seemingly multiple points of origin were separate and distinct and were not the result of the normal propagation of the fire.

- **Trailer** — An ignitable material used to spread fire. It usually leaves char or burn patterns and may be used with incendiary devices.

- **Incendiary devices** — Designed and used to start a fire. Most incendiary devices leave evidence of this existence, especially the metal parts of electrical or mechanical devices. More than one device may be used, and sometimes a faulty device can be found.

- **Ignitable liquid patterns** — Pronounced and irregular damage to a floor may indicate the use of an ignitable liquid. (This appearance can be created by carpet, concentrated fuel loads such as furniture, and ventilation.) Ignitable liquid flows to the lowest level possible; therefore, corners and the base of walls should be checked for an unconsumed accumulation.

- **Low levels of charring** — Low levels of charring indicate high amounts of heat at low levels and may indicate the introduction of ignitable liquids. However, the investigator must not jump to conclusions. Low-level burning may also be a result of the fire fighting tactics, flashover igniting existing fuels, or ventilation factors. Consider low-level burning along with all other data collected during the investigation before making any determinations.

- **Concrete spalling** — Spalling is caused by high heat liberating the moisture in the concrete, leaving its surface chipped and pitted. Newer concrete spalls more readily than does older concrete. Concrete spalling can occur as a result of burning ignitable liquid. If conditions are right, Class A or D combustibles may also cause concrete spalling. Many factors such as the type of aggregate and the age of the concrete affect spalling. Spalling may be caused by other situations such as the rapid cooling of heated concrete when fire streams are applied. The investigator must consider many factors when viewing spalled concrete.

- **Inverted V-patterns** — These may indicate the introduction of ignitable liquids. The inverted V-pattern is narrower at the apex (top) than it is at the base. These patterns are generally found on walls and originate at the floor. They often result from pooled ignitable liquids; however, other factors such as fall-down, ventilation, or fuel arrangement may also explain these findings. The investigator must rule out these other possibilities before concluding that the inverted V-patterns were caused by ignitable liquids.

- **Hourglass patterns** — Hourglass-shaped patterns are formed as a result of air movement from the side of the fire, cooling the wall surface where air enters the flame zone. These patterns may be the result of an ignitable liquid or any other material that releases heat at a high rate.

After the elimination of all accidental causes, the investigator can use the following "red flags" to assist in the investigation.

- Are contents removed or out of place?

- Is there an absence of personal items?

- Is there evidence of other crimes in the structure?

- Was the second fire in the same structure?

- Was there structural damage before the fire?

- Was the fire department access blocked?

- Did the fire occur at an unusual time of day?

- Were there signs of forced entry?

- Does the owner's/occupant's story match the circumstances?

- Did the fires occur during inclement weather causing a delayed response or providing a possible "excuse" for the fire?

 **Motives**

Adult firesetters are usually motivated by revenge, vandalism, profit (fraud), crime concealment, excitement (vanity), or extremism (terrorism). Fires are generally set in a location and manner unique to a specific motive. Identification of the motive provides valuable assistance in the identification of the party responsible for setting the fire.

### Revenge

Fires set because of personal or professional vendettas fall into the largest category of arson fires and account for fifty percent of the total arson problem. Generally, the victim is able to provide information regarding the suspect's identity. Personal property is often the target, and ignitable liquids are seldom used because most often the fire is not "mapped out," but rather it occurs without planning and in reaction to an incident. Normal targets are vehicles, storage rooms or outbuildings, and fences. Fires set to homes or businesses are often set to the exterior or through a broken window. Molotov cocktails or "firebombs" may also be employed. Revenge fires set in homes as a result of a spouse believed to be unfaithful are often set to that person's clothing or to the bed. Often a history of domestic disputes precedes the fire.

### Vandalism

Vandalism fires are most often set by two or more individuals (usually juveniles) for no apparent reason other than "just for kicks." Schools are prime targets, and other common locations include vacant buildings, trash containers, and vegetation. Forced entry to buildings is present, and property damage and graffiti are often done before the fire.

## Profit (Fraud)

Monetary gain is the primary motivator for this type of fire, and total destruction of property is the ultimate goal. The key to this fire is the desire to cause the most possible damage in the least possible time. For that reason, multiple fires and ignitable liquids are common. Holes broken in the walls or ceilings, trailers, etc., are often used to assist in the fire spread; time-delay ignition devices are not uncommon. Fires set by the property owners are often elaborate in nature and require a significant setup time. No other motive allows a firesetter unlimited access and time to the interior of a structure without fear of discovery.

Personal property that cannot be easily replaced and pets are often removed before the fire. The property owner is frequently absent from the building, and doors are found locked. Fraud fires are classically set because of poor financial status; however, the motivation for arson for profit may be quite abstract and as varied as the imagination of the firesetter. An example would include a profitable and well-established business set on fire for alternative fraud reasons; that is, a successful nightclub needs to be remodeled because of worn carpet and tobacco-smoked ceilings. If the owner were to close for the week required to complete the remodeling, revenue would be lost. If a fire was set to a business with adequate insurance, the overhead expenses, lost revenues, and remodeling fees would be paid through the business interruption insurance (if the fire were not properly investigated).

Other reasons for fraud fires do not involve the property owners. The fires are generally not elaborate and if set to the interior, require the firesetter to force entry. Examples of fires set for economic gain, not involving the owners or insured may include:

- Competitors to drive the victim out of business

- Contractors to secure a contract for rebuilding the loss

- Insurance agents to sell insurance to uninsured persons in the area

- Persons wishing to devalue the property so they can purchase it at a lower price

- Firefighters to obtain overtime or call-out pay

## Crime Concealment

Arson used as a tool to destroy evidence of another crime is most generally associated with burglary, homicide, and embezzlement. The attempt to cover a burglary is most common with the fire set at the location where evidence, such as fingerprints or blood, was believed to have been left. Most often the location is at the point of entry or where an item has been removed. The fire is generally set with combustibles on hand and rarely involves ignitable liquids as a burglar usually enters a structure with the intent to steal and not to set a fire. The fire is set after entry and after it is decided that incriminating evidence was "left behind."

Homicide concealment fires, however, often involve the use of ignitable liquids in an attempt to destroy the body and therefore evidence of the manner and cause of death. These fires are generally set on and around the body.

Embezzlement fires are set to erase or destroy a "paper trail"; therefore, the paperwork and surrounding area are the origin for the fire. Often the paperwork itself is used as the fuel with ignitable liquids sometimes used to assist in the destruction of the documents.

## Excitement (Vanity)

Excitement and action accompany a fire and the ability to create a situation requiring the response of the fire service and law enforcement provide some people with a feeling of empowerment over society. The spur-of-the-moment fires, however, develop as a recognizable pattern over a period of time. Examples of pattern development include:

- **Dates and day of the week.** Paydays, normal work days, or days spent consuming alcohol are believed to help stimulate these individuals in firesetting.

- **Time of day.** The time of day or night may correspond with travel to and from work or other activity. Most excitement fires are set during the hours of darkness.

- **Type of structure.** The arsonist is often consciously or subconsciously attracted to a certain type of structure — for example, schools, churches, vacant structures, etc.

- **How the fire is set.** The arsonist rarely plans to set a fire; therefore, combustibles on hand are most often used. They often become "comfortable" with a certain method and tend to stay with the method that has worked in the past.

- **Where the fire is set.** The arsonist tends to set these fires in similar locations (for example, under a crawl space) as prior fires at that location resulted in the required emergency response and lack of detection.

Arsonists who seek recognition or wish to be viewed as heroes may set and "discover" fires. These individuals are always present at the fire scene and often attempt to assist in fire fighting activities. They may be from any background; however, it has been noted that some have been employed as security guards, volunteer firefighters, and reserve law enforcement officers. These same individuals may often be seen at multiple fire scenes, and if their presence is observed, the investigator should check their background for past examples of firesetting behavior.

## Extremism (Terrorism)

Social protest by an individual or group may target a government, ethnic, or religious group or a facility that operates in opposition to their "cause." Fires or explosions are carried out with the intent to advertise or advance the arsonist's purpose. Although the arsonist wishes his individual identity to remain unknown, it is important that their group or "cause" be identified as the responsible party. Graffiti or signs may be left at the scene, and phone calls or letters to the press are common. Fires and explosives are most often set to the exterior of buildings or are propelled (as is a Molotov cocktail, for example) into the interior through broken windows or doorways.

## Pyromania

Pyromania has not been included as a motive because it is a mental state and a recognized psychological disorder. True pyromaniacs are few in number and set fires as a release of tension or in response to "voices" from within or in the form of an imaginary person or animal. These fires are seldom set with ignitable liquids and are often set to paper products in vehicles, in alleys, or behind

buildings. Over a short period of time it is common for this type of individual to set multiple small fires within several blocks of each other.

 **Opportunity**

The evaluation and correlation of various evidence show who had the opportunity to commit the crime. Evidence to be considered may include the following:

- Personal items found at the scene
- Building security
- Latent prints
- Impressions
- Phone records
- Witness statements
- Alarm system printout
- Security or surveillance cameras
- Traffic citations

### Financial Analysis

The investigator reviews and evaluates the current financial condition of an individual or a business. The evaluation starts with a review of the most basic factors. These factors are:

- Assets
- Liabilities

Any anomalies that emerge from this review should be investigated further. For example, any extraordinary liabilities that the individual or business may have, such as business or personal loans coming due, may be motivating factors. Likewise, a pattern of diminishing assets such as a series of business losses may also be a motivating factor.

 **Conclusion**

Once a thorough examination of a fire scene has been completed, the investigator moves into a phase of the investigation that involves the organization and analysis of the available information with the ultimate objective of determining the origin and cause of the fire. When it is determined that a fire was intentionally set, the fire investigator may also be involved in determining who was responsible for the fire.

# Chapter 16

# Presenting Investigative Findings

Typically, the investigator's job does not cease with the termination of investigation procedures at the fire scene. Most investigators will be responsible for summarizing and presenting their investigative findings in some type of formal manner. Depending on the circumstances and local agency policies, investigators may find themselves handling media inquiries, preparing written reports, or providing testimony in court proceedings. This chapter discusses the basics that investigators need to know about each of the responsibilities.

##  Media Relations

Good media relations are vitally important to the public image and credibility of the fire investigator and the agency he represents. Taking a professional and proactive approach when dealing with the media benefits everyone involved: the media, the investigator and his agency, and the public.

Interviews and news releases are the two primary ways in which interaction between the media and the fire investigator may occur. This section highlights each of these.

### Interviews

A media interview may be planned and prearranged, or it could be a spur-of-the-moment interview on the fire scene. In any case, the investigator must be prepared for this possibility and represent the department or company well as a spokesperson.

Speaking with or being interviewed by the media can be a challenge. The media has the ability to either support or complicate the work of the inves-

tigation. Investigators should work with the media, be honest and forthright, and follow their departmental or company policies and procedures. Some departments have a designated information officer (PIO) assigned to answer media questions on large incidents. If the PIO is on the scene, investigators should politely direct questions and interview requests to the PIO. If a PIO is not assigned or available, investigators should follow established policies and procedures when being interviewed. The following guidelines will be helpful to investigators who are assigned to give an interview:

- There is no such thing as "off the record." Anything said to a reporter can be quoted. Do not be misled by the friendly reporter who says "Just between us..."

- Beware when asked leading questions. Sometimes reporters use this tactic to get the answer they want. Listen to the questions carefully and thoughtfully and answer "yes" or "that's correct" only if the facts in the questions are 100 percent accurate and no inaccurate conclusions have been drawn.

- Avoid getting into disagreements or becoming defensive with reporters.

- Defensiveness may suggest that information is being withheld, even if it is not.

- Do not be led into answering questions beyond your area of expertise.

- Refer such questions to those who have the necessary information, or offer to find the answer — and *always* follow through.

- Avoid using esoteric fire service terminology. If a technical term must be used, explain its meaning at the time it is used.

- Honesty is the best policy. Be as frank and open as possible without divulging confidential information about victims' identities, possible fire cause, etc.

- Do not answer "What if..." questions. Do not answer hypothetical questions, explaining that you are not prepared to speculate.

- Listen for false or misleading information in reporters' questions. If a question contains false information, politely discount the misinformation and provide accurate information.

- Beware of the forced-choice question. If either way of answering the question would be inaccurate, then answer with a separate, factual response. Be tactful, but refute the false information.

- Do not volunteer information, especially if it is speculative. For example, to prematurely suggest to the media that a fire is possibly of electrical origin could have an adverse affect if an arson case is later developed from that fire.

- Be prepared. Rehearse your interview technique, and try to improve delivery.

## News Releases

The *news release*, also known as a press release, is used to provide the media with information in a ready-to-use news story format. As an investigator, you may be responsible for preparing a news release that summarizes the current investigation on which you may be working. The cause, fire loss, and details of the prosecution of arsonists should be included in the investigation news release. It is important to remember that when releasing information regarding an investigation, you must consider the legal and ongoing investigation issues associated with the incident. You may also be asked to provide information regarding the investigation to the public information officer or company officer who may be preparing a news release on the incident.

### Parts of the News Release

The news release has two parts: the heading and the body. The *heading* generally identifies the sub-ject of the news release, the date the news release is being disseminated, the point of contact at the fire department, and when the news release can be used. The *body* or content of the news release contains the actual story that you want to get to the public. The story is written in the same style as a newspaper story. The story should be concise and easy to read. Also, the news release should be double-spaced and on department letterhead. It must be free of grammar and spelling errors.

The story begins with a title or headline, which summarizes the meaning of the story. The first paragraph of the story is the most important and should hook the reader's interest. This paragraph should summarize the "who, what, when, where, and why" of the story. The reader should be able to read the first paragraph and get the basic idea about the story. Why? If the reader just reads the first paragraph, he or she will get the basics of the story. A rule used by newspaper journalists is that the most important parts of a story are at the beginning. As the story in a newspaper progresses, the importance of the information decreases. This allows the editor to cut the length of the story from the bottom without compromising the story. If your first paragraph contains the most important part of the story in a summary manner, the reader will get the information even if the rest of the story is cut by the editor or never read.

### News Release Tips

- Keep them short. Provide only the most important information to keep the story interesting. If the reporter wants more specific information, he or she can contact you and get a follow-up interview.

- Get to know the needs and preferences of your local media, and then compose your releases in that fashion.

- When you can, include quotes from decision makers and those involved in the incidents and the stories. The public loves to read about the human interest part of emergency services.

- Only put information in your news release that you know — without a doubt — is accurate and factual. Once you have sent it out, it is very difficult to retrieve and change.

- Ask at least one other person to proofread the news release before disseminating it to the media.

- When sending out a news release, make sure that it goes to all the local media. Failure to send it to all media may be perceived as preferential treatment and can cause mistrust and hurt relationships with the media.

- Remember that news does have a usable life. Old news is seldom used. You need to get your news release to the media in time for their news deadline. The best way to do this is by fax or e-mail. If you live in a small community, you may even deliver it yourself. Avoid sending it by mail as it may not get there in time to be up-to-date news.

 ## Report Writing for the Fire Investigator

In order to certify to NFPA 1033, fire investigators must be able to prepare a written report that documents the investigation they performed. However, in the actual performance of their duties in the field, whether or not the investigators are required to file a formal written report will be a matter of agency or company policy. This policy may also dictate the style of the report and the content that is required. From the outset of the incident, the investigator should clarify whether a written report will be required by the agency employing him. It is important for the investigator to keep in mind that even though a written report may not be requested at the beginning of the fire investigation, one may be required at a later date.

Reports are generally required of investigators employed by public agencies, such as fire departments. The reports are used for arson prosecutions, development of crime statistics, identification of fire-related trends (that is, to establish that a specific brand of appliance continually causes a fire), and many other purposes. Public agency reports are often considered public record and may be read by victims, supervisors, attorneys, insurance representatives, and the public at large.

Reports may or may not be required of investigators employed in the private sector. This necessity is generally controlled by company policy, the entity on whose behalf the fire is being investigated, or

the judicial system. Investigations are conducted in the same thorough manner as indicated throughout this manual, whether or not a report is required.

## The Written Report

Local policy dictates to whom, if anyone, the report will be addressed. This could be the fire chief, fire marshal, law enforcement personnel, or any other person as directed by standard operating procedures. In some cases, the report will simply be a narrative that is not addressed to anyone. In either case, the report should be viewed as a "legacy" of the investigation.

The primary purpose for producing a report in writing is to relay to the reader all the usable information obtained during your investigation. The report should provide the intended audience with only the pertinent information obtained during the fire investigation and all information that is necessary to continue the investigation. This information should be relayed in a concise and understandable manner. The information to be included in the report is somewhat subjective; however, the investigator should always remain neutral when determining what information should be included. If the information obtained is pertinent, report the data whether it vindicates or implicates the suspect.

The purpose of the report is not to build a case but rather to report the facts as you know them. Supposition and assumptions are not appropriate in an investigative fire report. Your opinion, based solely on the cumulative facts obtained during the investigation, should be presented in a manner that is coherent, concise, and understandable. When required, a thorough and professional investigative report may increase the chance of a successful disposition of the case.

Many word processing programs are available to assist you in producing a professional report, free of spelling errors. However, this word processing program should not completely take the place of another competent individual proofing your report. A segment of the report may be found confusing and require rewording or rewriting.

A table of contents and an introduction are two basic parts of a written report that are typically included.

## Table of Contents

Depending on the size and scope of your report, it may be necessary to begin with a table of contents to assist the reader through the labyrinth of information. The table of contents provides a road map to the body of the report. It also highlights the presence and location of attachments to the report such as law enforcement reports, fire department reports, medical examiner's report, laboratory analysis, and interviews.

## Introduction/Preface

This portion of the report may contain a short synopsis of the facts and conclusions that follow in more detail in the main body of the report. In some writing styles, this is commonly referred to as an *executive summary*. It may summarize details such as the date and time of the incident, the reporting party, and the fire cause determination.

Some investigators include in this portion of the report a detailed explanation of how this assignment was received and who instructed or requested the investigation. The investigator should provide the scope of the requested inquiry and note the chronological order of actions taken during the investigation.

# Types of Information Used

The information in the following section highlights some of the types of information that are commonly used in fire investigation reports. This is to be used only as a guide for the fire investigator who is learning to write an investigative report. The order in which the information is presented may be altered to reflect the policies of your company or agency.

## Fire Scene Description

It is important to provide a thorough description of the property involved in the fire loss. When dealing with a structural fire, this information may include items such as the following:

- Year of construction

- Height or number of stories

- Occupancy usage

- Configuration, including the direction the structure faces

- Type of foundation

- Exterior veneer

- Style of roof construction and the covering

- Type of windows

It is important to include any information about the building construction that assists the reader in understanding the report, the path of the fire spread, and the origin and cause for this loss. Using the 2nd edition of the IFSTA **Building Construction Related to the Fire Service** manual may be helpful in identifying many of these building characteristics.

If the incident being reported on is not a structural fire, the type of information presented in this section will obviously be slightly different. For example, when reporting on a vehicle fire, it is important to include information such as the year/make/model of the automobile and its location when the fire occurred. When reporting on a wildfire, it is necessary to address terrain features, types of ground cover, and other pertinent information. Based on the particular incident being investigated, investigators must use their judgment to determine the types of information that are important to note.

## The Investigative Process

The next major portion of the report should include a detailed description of the process used to perform the investigation and a brief explanation describing the method used in processing the fire scene. Whatever the investigative method used, the report should follow the same order as the investigation.

The investigator should include the methods and procedures that he used to collect evidence and identify in a clear and accurate manner the evidence that was seized. The person or persons who collected the evidence should be listed in the report — this begins the chain of custody for the evidence seized. At the time of writing the report, the investigator should note the location where the evidence is stored.

## Exterior Documentation Phase

In this section of the report, the investigator explains each of his actions during the exterior documentation phase of conducting the fire investigation. He should clearly explain where the examination began and in which direction it was conducted. Some of the items that may be pertinent and therefore should be noted during this part of the investigation and report include:

- Tire tracks or unusual items found at the scene

- Locked gates or other important items affecting entry to the premises

- Fire hydrants, fire department connections, and other private fire protection equipment and their condition

- Condition of the lawn and area surrounding the structure

- For animals normally housed outside, the presence or absence of shelter and bowls for food and water

- Information from the public utilities. (**NOTE:** If applicable, each of the utilities [gas and electric] should be eliminated as providing the direct heat or fuel for the incident.)

- Fire patterns and the area of greatest fire damage

- Windows or doors opened during the fire, which may help in explaining the directional spread of the fire

- Information regarding weather that may have affected the fire suppression activities as well as flame spread, including whether lightning was reported in the area prior to the discovery of the fire

### *Door Assembly Information*

The entry doors at any fire loss have the potential for offering some very important information. Information that confirms or denies witness statements may be provided in the report. Reporting on the relationship between the security positions and the forced entry marks or the lack thereof should be relayed during this portion of the report. Any low-level burning that would not be attributed to ventilation or any other known phenomenon should be stated.

### *Interior Documentation*

This section should include other observations that are relevant to the fire investigation. The investigator should provide a brief description of the initial point of scene processing. The report should include a description of the general condition of the interior and the interior finishes of the structure. The investigator should state whether contents found inside the loss site appeared to be normal or abnormal and whether hygiene items, photographs, and food products were found.

The next step is to identify the area of origin with any explanations necessary to qualify this determination. A detailed description of the directional flame and heat spread that occurred within the structure should be included. Secondary burning —fall down or exposure of additional fuel—should also be explained. This lays the foundation for identifying the area of origin.

After identifying the area of origin, the investigator should include a list of the potential heat sources located within this area and justification for eliminating each of these possible sources of ignition. The identity of the material first ignited should likewise be provided. The investigator's opinion of the cause for the fire loss should be presented with all conclusive data that supports the opinion.

The report should document the causal factors surrounding the ignition for this fire. The explanation as to how the heat and fuel came together to produce the fire needs to be presented in an understandable manner.

If the cause for the fire has not been determined, this information should be provided as an undetermined cause. Examinations still being conducted that may affect the investigator's position as to the cause for the fire should be noted with a full explanation.

## Photographs

In some cases, photographs may be included with an investigative report. As with the report, your company or agency policy dictates what information is present. Often the minimum information provided for each photograph includes a numeric identifier for the roll and exposure, the viewing direction, and a brief explanation of what the photograph depicts.

## Diagrams

Demonstrative evidence is a vital part of any fire investigation, especially when presented in a legal setting. Diagrams are extremely important and should be produced with great care and accuracy. Examples of diagrams include room identifications and interior measurements, contents, appliances, utilities, directional flame spread, and area of origin. Many inexpensive computer programs are available that will produce a quality diagram. Time spent in producing this aspect of the report will help to clarify the evidence being presented. If evidence has been seized, a diagram may be produced to show the location and identity of the items seized.

## Interviews

As the experienced fire investigator knows, the interviews conducted are of primary importance in the successful disposition of the investigation. Incorporating the information received into a report form is equally important. Some interviews will already be in written form; however, some will require the investigator to produce them in an understandable manner. Great care should be given to make sure that the interview information placed in the report accurately reflects what the interviewed party stated. The investigator should not leave out information that may assist the subject, remembering always that the investigator is a neutral party who is seeking the truth.

If the interview was recorded, a verbatim transcript of the interview should be included. All valuable information should be outlined in a summary format in order that the reader can view and, if necessary, retrieve the data from the transcript.

### Fire Incident Report

The incident commander or the assigned representative will normally fill out the fire incident report. These reports are usually standardized and should become an attachment to the investigative report.

## Conclusion or Opinion

The final portion of the formal body of the report is the section where the investigator highlights each of the conclusions and opinions that are relevant to the incident. These opinions must be based solely on the facts of the incident. Using this section to state "gut instincts" or unsubstantiated hypotheses is unprofessional and could be damaging to

the case. For more information on investigative report writing, see NFPA 906, *Guide for Fire Incident Field Notes*.

 ## Legal Proceedings

All investigations should be conducted under the premise that the investigator will be subpoenaed to provide sworn testimony in a court of law. The case may be criminal in nature in the event that a fire cause is determined to be incendiary or the result of an intentional act. The case may be civil involving code violations, alleged product defects, or other noncriminal actions. Many persons are understandably apprehensive when called to testify in court as the judicial system may be viewed as complicated and mysterious. This section focuses on simplifying and explaining the litigation process.

### Civil Proceedings

A *civil proceeding* may be the result of a noncriminal violation but is more often a result of one party (one or more people or a large company) that files suit against another party. This suit may allege a breach of contract or be for any other reason, which is known as a *tort*. The following is an example of a civil or tort action:

> John Doe purchases a new electric heater from a local retail mall. Several years later while in use, this heater suddenly malfunctions into a flaming state. This causes damage to the home, as well as burn injuries to Mr. Doe as he attempts to extinguish the fire. John Doe later "files suit" against ABC Corporation, manufacturer of the heater, and XYZ Corporation, the retail store that sold him the appliance. The suit is filed, alleging that the ABC Corporation manufactured an unsafe product and that XYZ Corporation sold that unsafe product.

The party filing the suit (*plaintiff*) is required to establish through a preponderance (more likely than not) of evidence that the party sued (*defendant*) did, in fact, commit the allegation. Specific time frames are established to govern the "discovery" process. *Discovery* is the means by which the plaintiff (one party) obtains information from the opposing party (defendant) to prove its allegation regarding the information and/or evidence that it possesses. The defendant can also use the discovery process to show that he or she is not responsible for the allegation. These time frames vary among jurisdictions, and specific information regarding this area should be obtained from local attorneys.

Rules governing what disclosure must be made available also vary among jurisdictions. For example, some state jurisdictions do not require a party to provide reports from their expert witness(es). On the other hand, federal district courts require a written report from each testifying expert involved in the action.

Most jurisdictions provide rules indicating that each party to the suit may depose the opposing party's witnesses. A *deposition* is a process during which the witness answers questions under oath posed by the attorneys for each party. The deposition takes place without a judge or jury, and questions and answers are recorded by a court reporter. The questions typically regard the facts surrounding the allegation. However, if the witness is an expert, the questions are often based on his or her opinion relating to the allegation. Each party uses the deposition testimony to prepare for the trial. It is also used in trial if the witness is not available or if the witness provides testimony that is different than that given during the deposition. The process of showing the judge or jury that a witness has changed his or her opinion or prior testimony is known as *impeachment.*

During the trial phase of the civil proceeding, the judge makes determinations based on law as to what evidence is presented to the jury. The judge may exclude all or part of a person's testimony or items of physical evidence based on his or her interpretation of the law. In some cases, the judge will also serve as the determinant of the outcome of the case. This is known as a *bench trial,* and both parties in the case must mutually agree upon this. Otherwise, the decision will be up to a jury in a *jury trial.* Most jurisdictions do not require a unanimous decision from the jury, and the number of jurors impaneled and the number required for a verdict vary by jurisdiction.

As the plaintiff has the "burden of proof," the plaintiff's evidence must be presented first. The defense then provides its evidence, and the matter is presented to the trier of fact (judge or jury) for a determination or verdict. As previously stated, the verdict is established on a preponderance or majority (51 percent) basis. The final outcome of the case may be a full judgment for either side or a partial judgment that is split between the two parties.

## Criminal Proceedings

Criminal proceedings vary according to jurisdiction, and the prosecuting attorney's office should be consulted for complete information regarding the local process. Criminal proceedings are most often initiated as a result of action taken in the form of a "probable cause" arrest, service of an arrest warrant, or grand jury indictment. This action is the result of an investigation establishing that a crime has occurred and that sufficient probable cause is present to believe that the arrested or indicted party committed the crime. *Probable cause,* simply stated, is having enough information or facts to believe that it is probable (more likely than not) that the party is responsible. This level of proof is not to be confused with the level necessary for a finding of guilt by the trier of fact. The potential punishment a party could receive if found guilty (responsible) most often dictates whether the crime is a misdemeanor or a felony and how the case proceeds through the justice system.

## Classifications of Criminal Violations

In most jurisdictions, there are three types of criminal violations with which the investigator should be familiar:

- Summary offenses
- Misdemeanors
- Felonies

### Summary Offense

A *summary offense* is the lowest form of offense in most legal systems. These are minor infractions of laws or local ordinances. The person who commits a summary offense is typically given a written citation of the violation on the spot when the violation occurs. He then has the choice of simply paying the fine that accompanies the citation or challenging the citation before a judge. The judge may uphold the citation or dismiss it. The most common forms of summary offenses include speeding or parking violations and similar offenses. In some jurisdictions, unauthorized controlled burns (trash, leaves, etc.) are considered summary offenses, and the fire investigator may be involved in writing the citations. It is important to clarify under local law whether the fire investigator has the legal authority to write a citation. If not, he may need to call a police officer to the scene to perform the task.

### Misdemeanor

A *misdemeanor* is generally viewed as a "lesser" crime and is punishable by a fine or (generally) a term of less than two years in jail or prison. In the event of a misdemeanor, the arrestee is given an initial appearance, during which he or she is allowed to "plead" guilty or not guilty. If the plea is guilty, a sentence may be imposed or levied in the form of a fine, incarceration, or both. If the subject pleads not guilty, a trial before a judge or jury is set.

### Felony

A *felony* is generally viewed as a "serious" crime and is punishable by a fine, prison, and/or death, depending on the severity of the felony and the jurisdiction. Like the misdemeanor, the arrestee is given an initial appearance and the option of pleading guilty or not guilty. If the plea is guilty, a sentence may be imposed at that time or after a study outlining the circumstances and the arrestee's past actions has been completed. If the plea is not guilty, a preliminary hearing (also known as a "probable cause hearing") date is set. A person indicted by a grand jury and pleading not guilty at his initial appearance is immediately "bound over for trial" and the preliminary hearing stage is bypassed.

##  Court Proceedings

Fire investigators should be familiar with the various types of court proceedings with which they may become involved. This section highlights some of the more common ones.

## Preliminary Hearings

A preliminary hearing takes place before a judge. The prosecutor's office is required to establish that a crime has been committed and that it is more likely than not that the arrested party has committed the offense specified. If sufficient evidence is presented, the judge may "find in favor" of the prosecutor and "remand or bind the defendant (arrestee) over for trial." Most jurisdictions will require a fire origin and cause expert to establish that the fire was a result of a crime (typically intentionally and maliciously set) prior to "a finding of probable cause."

## Felony Trial

In the event of a felony trial, the defendant (person charged with the crime) has the constitutional right to a trial before a jury. The defendant sometimes waives this "right," and the trial takes place with only the judge (bench trial) making the determination as to the defendant's guilt or innocence. The burden of proof is quite different at the criminal trial stage and requires the prosecutor's office to present evidence sufficient to convince the judge/jury of fact to base the defendant's guilt on a finding of "beyond a reasonable doubt." A reasonable doubt standard does not require a 100 percent certainty; however, it does require a level of proof beyond that which a "reasonable" person would have after hearing all the evidence.

As the prosecutors have the burden of proving that a crime was committed and that the person standing trial was the one responsible for that crime, they present evidence to the trier of fact first. The defendant and defense attorneys then present evidence in support of their not guilty position after the prosecutor finishes or "rests." Witnesses are often called upon to testify to facts indicating that a crime has taken place (corpus delicti) or that a person is responsible for the crime.

##  Testimony

*Testimony* is simply an affirmation or declaration made under oath, typically in a court of law. All verbal information provided in a trial to the trier of fact is in the form of sworn or affirmed testimony in which the witness promises to tell the truth under penalty of perjury. The type of information presented by the witness varies. It is generally divided into two categories: fact witness and expert witness.

## Fact Witness

A *fact witness* is a witness who answers questions posed by an attorney regarding what he or she saw, heard, touched, smelled, or tasted. The testimony presents information to the judge or jury regarding "facts" without the witness being allowed to provide his or her own interpretation of those facts. An example of fact testimony (also known as "lay testimony") would be a witness's description of seeing a fire in the corner of the living room on the couch as seen through the front window. Typically, that witness would not be allowed to provide further testimony; that is, the length of time the fire was burning, what caused the fire to occur at that location, etc. This further type of testimony would be classified as expert testimony.

## Expert Testimony

In contrast to the fact or lay witness, an expert witness is allowed at the discretion of the court to provide opinion testimony; for example, the point of origin was the living room couch and the ignition source was electrical as a result of the couch leg sitting on the lamp extension cord. An *expert witness* is generally defined as a person with sufficient skill, knowledge, or experience in a given field so as to be capable of drawing inferences or reaching conclusions or opinions that an average person would not be competent to reach. The expert's opinion aids the judge or jury in understanding the facts at issue and assists them in the search for the truth, also known as a *verdict*.

The evidence that the investigator uses and upon which he bases his expert opinion typically must be relevant and reliable, thereby allowing it to be ad-missible. The question as to whether that evidence is relevant and/or reliable is at the sole discretion of the trial judge and, therefore, may be open to interpretation from court to court or between jurisdictions. United States Supreme Court decisions including the "Daubert Decision" (Daubert, et. ux, etc., et. al., Petitioners v. Merrell Dow Pharmaceuticals, Inc., 509 U.S. 5709 [1993]) and the Carmichael Decision (Kumho Tire Company, Ltd. v. Patrick Carmichael, et. ux, etc., et. al., 119 S. Ct. 1167, March 23, 1999) have particular relevance to the fire investigator. These Supreme Court decisions are federal court decisions; however, these are widely accepted as a standard within state level and lower court levels. Through these decisions, the trial court judge has the responsibility as a gatekeeper to allow or disallow expert testimony utilizing accepted criteria. That criteria may include the reliability of the methodology used by the investigator as well as the general acceptance within the fire investigation community of that methodology.

 **Conclusion**

Starting with the initial call to the fire scene, the investigator should assume that the investigative report will be scrutinized in a court of law as part of civil litigation or criminal prosecution. The investigator should also anticipate being called to testify as an expert witness. Therefore, the investigator must be careful to use accepted methodology at every step in the investigative process. In addition, the investigator must know the requirements and limitations imposed by local, state/provincial, and federal law and scrupulously meet those requirements and operate within those limitations.

# Educational Resources for Arson and Fire Prevention

The insurance industry has developed effective tools for educating the public. The resources listed below can provide communities, organizations, and public safety agencies with literature, planning manuals, audiovisual materials, media campaign materials, and, at times, seed money for programs.

Aetna Life and Casualty
Arson and Fraud Unit
151 Farmington Avenue
Hartford, CT 06156
Contact: John L. Swedo, Asst. Vice President
800-323-8648

Alliance of American Insurers
1501 Woodfield Road, Suite 400 West
Schaumburg, IL 60173-4980
Contact: Dean M. Moffitt, Vice President, Personal
   Lines Division
708-330-8526
708-330-8602 (Fax)

Allstate Insurance Company
Allstate Plaza F3
Northbrook, IL 60062
Contact: David B. Warstler, Senior Corporate Relations
   Manager
708-402-2908

Factory Mutual Engineering and Research Corporation
1151 Boston Providence Turnpike
P.O. Box 9102
Norwood, MA 02062
Contact: Donald C. Garner, Arson Coordinator
617-762-4300

ITT Hartford Insurance Group
Hartford Plaza
Hartford, CT 06115
Contact: Melissa H. Engel
203-547-4711

Insurance Committee for Arson Control
110 William Street
New York, NY 10038
Contact: Richard Gilman, Executive Director
212-669-9245
212-732-1916 (Fax)

Insurance Information Institute
110 William Street
New York, NY 10038
Contact: Gordon C. Stewart
212-669-2900

New York Property Insurance Underwriters
   Association, Inc.
110 William Street, 4th Floor
New York, NY 10038
Contact: Kenneth Lang, Asst. Vice President, Claims
212-208-9813

Professional Insurance Agents
400 N. Washington Street
Alexandria, VA 22314
Contact: James Quiggle, Director, Public Relations
703-836-9340

SAFECO Insurance Companies
Safeco Plaza
Seattle, WA 98185
Contact: Gordon C. Hamilton, Vice President of Public
   Relations
206-545-5705

St. Paul Fire and Marine Insurance Company
385 Washington Street
St. Paul, MN 55102
Contact: Karen Himley, Vice President,
   Communications
612-221-7911

State Farm Fire and Casualty Company
112 E. Washington Street
Bloomington, IL 61701
Contact: David Stuart, Claim Consultant
309-766-2983

Western Insurance Information Service
Colorado Advisory Committee on Arson Prevention
6565 S. Dayton Street, Suite 2400
Englewood, CO 80111
Contact: Shannon Kelly
303-790-0216

# Federal Constitutional Search and Seizure Issues in Fire Scene Investigations

## Introduction

This memorandum summarizes the state of the law with respect to federal constitutional search and seizure issues in fire scene investigations. It does not address search and seizure issues with respect to State law, which might afford more protection to property owners and place greater restraint on fire investigators than does the federal constitution. See *Mills v. Rogers*, 457 U.S. 291, 303 (1982). However, because the federal constitution provides the minimum level of search and seizure protection below which the States may not stray, this summarization provides the minimum level of protection generally applicable in the United States. Id.

## The Law

The Supreme Court outlined the Fourth Amendment search and seizure protections applicable to fire scene investigations in the cases of *Michigan v. Tyler*, 436 U.S. 499 (1978), and *Michigan v. Clifford*, 464 U.S. 287 (1984). In these cases, the Court held that the conduct of fire scene investigations fall within the strictures of the Fourth Amendment's protection against unreasonable searches and seizures. *Tyler*, 436 U.S. 499, 506 (1978). Accordingly, the general rule is that fire investigators may not make a nonconsensual entry onto private property without a duly authorized search warrant. *Tyler*, 436 U.S. at 506. However, the Court granted fire investigators some latitude to conduct warrantless searches of fire scenes by allowing fire officials to remain on private property for a reasonable time after the fire has been extinguished, and by allowing the use of administrative search warrants for routine fire investigations.

The Court has recognized that "[a] burning building clearly presents an exigency of sufficient proportions to render a warrantless entry 'reasonable.'" Id. at 509. Because the firefighting function encompasses "not only ... extinguishing fires, but with finding their causes," fire officials can remain on private property for a "reasonable time" after the fire has been extinguished to determine the origin of the blaze. Id. at 510. However, what constitutes a "reasonable" time is a very fact specific determination that may vary with the circumstances of each case. Id. at 510 n. 6.

In order to obtain a search warrant for the purpose of conducting a routine investigation into the origin of an unexplained fire, fire officials need only establish the level of probable cause required for administrative searches. *Michigan v. Clifford*, 464 U.S. 287, 294 (1984). This level of probable cause is established upon a showing that "reasonable legislative or administrative standards for conducting an ... inspection are satisfied with respect to a particular dwelling." *Camara v. Municipal* Court, 387 U.S. 523, 538. As applied in the fire investigation context, this standard requires fire investigators to "show only that a fire of undetermined origin has occurred on the premises, that the scope of the proposed search is reasonable and will not intrude unnecessarily on the fire victim's privacy, and that the search will be executed at a reasonable and convenient time." *Michigan v. Clifford*, 464 U.S. 287, 294 (1984). However, to search for evidence of arson, officials must establish the traditional level of probable cause required of searches for evidence of crime. *Tyler*, 436 U.S. at 512. Criminal search warrants require a showing that the facts and circumstances of which investigators have "reasonably trustworthy information" are sufficient to cause a person of "reasonable caution" to believe that an offense has been committed. *Draper v. United States*, 358 U.S. 307, 313 (1959).

## Supreme Court Cases
### *Michigan v. Tyler, 436 U.S. 499 (1978)*

In *Tyler*, a fire department responding to a midnight fire in a furniture store was able to control the blaze by approximately 2:00 a.m. As the firefighters were "watering down smoldering embers," the Fire Chief arrived and entered the smoking building to examine two containers of flammable liquid that had been found in the store. Believing the fire "could possibly have been arson," the Fire Chief called a detective who arrived at approximately 3:30 a.m. *Tyler* at 502. The detective entered the building to begin his investigation, but was forced to abandon the effort because of steam and smoke in the building. The firefighters completely extinguished the fire and departed by 4:00 a.m. At approximately 8:00 a.m., the Chief returned with the Assistant Chief for a brief examination of the scene. The Assistant Chief returned again at 9:00 a.m. with the detective. They found and seized evidence of arson, including what appeared to be a fuse trail burned into the carpet. Several weeks later, other investigators returned to the scene and found further evidence of arson. Although the building owners did not object to theses entries and seizures at the time, they objected when the evidence was introduced against them at trial. The Michigan Supreme Court held that all the entries made by fire investigators after the flames had been extinguished were illegal warrantless searches, ruled inadmissible any evidence found as a result of these entries, and reversed the convictions of the defendants.

The U.S. Supreme Court held that a property owner's reasonable expectation of privacy is not destroyed simply because his property has been damaged by fire. *Tyler*, 436 U.S. at 505. Having established that an expectation of privacy remained, the property owner was protected by the Fourth Amendment's prohibition on unreasonable searches and seizures. Id. at 506. However, the Court also recognized that there are circumstances in which official action is compelled, but there is no time to obtain a search warrant. Such compelling circumstances can arise in either a criminal investigation or administrative inspection context. Id. at 509.

The Court held that a "burning building clearly presents an exigency of sufficient proportions to render a warrantless entry [to put out the blaze] 'reasonable.' . . . [Once] in the building for this purpose, firefighters may seize evidence of arson that is in plain view." Id. at 509. In addition, because "[fire officials are charged not only with extinguishing fires, but with finding their causes ... [they] need no warrant to remain in a building for a reasonable time to investigate the cause of a blaze after it has been extinguished." Id. at 510. What constitutes "a reasonable time" is a very fact specific determination. The Court commented in a footnote that a number of factors go into the "reasonableness" determination, including the type of structure, the size of the fire, as well as the individual's reasonable expectation of privacy. Id. at 510 n. 6. In this case, the Court decided that the warrantless re-entries of the property the following morning were justified as "no more than an actual continuation of the first [entry], and the lack of a warrant did not invalidate the resulting seizure of evidence." Id. at 511. However, the entries onto the property weeks later were found to be "clearly detached from the initial exigency." Id. at 511. Because investigators obtained neither administrative nor criminal search warrants for these later entries, the evidence obtained was inadmissible at trial. Id. at 511.

### *Michigan v. Clifford, 464 U.S. 287 (1984)*

After an early morning fire at the Cliffords' residence was extinguished at 7:04 a.m., all police and firefighters left the scene. Arson investigators, having been notified that the fire was suspicious, arrived at the scene five hours later. They found a work crew busy, on the vacationing Cliffords' instructions, securing the house and pumping water from the basement. While waiting for the water to be pumped, the investigators found a fuel can in the driveway which had been removed by firefighters and seized it as evidence. Although the investigators knew that the Cliffords' had given instructions to secure the house, they entered the residence without consent or a warrant and conducted a thorough search. In the basement they found two more fuel cans and a crock pot attached to a timer, which they also seized as evidence. They then continued their search into the living areas of the residence and found further evidence of arson. Before trial, the Cliffords' moved to exclude all evidence that was seized during this search of their residence.

The Supreme Court rejected the State's assertion that all postfire administrative searches should be exempt from warrant requirements, and affirmed the principle that, absent exceptional circumstances, all nonconsensual searches require warrants. *Clifford,* 464 U.S. at 291-92. The Court also rejected the State's suggestion that the *Tyler* principle allowed the warrantless postfire search as a continuation of the entry made by the firefighters. The Court distinguished the two cases not on legal principle, but on the particular facts of each case.

In *Tyler,* investigators were forced to call off the initial attempt at investigation because of smoke and darkness, but resumed the search as soon as practicable. *Tyler,* at 296. The Court found that the Clifford's efforts to secure their residence during the time between the departure of the firefighters and the arrival of the arson investigators separated the entry of the firefighters and the entry of the investigators into two different events and precluded considering the investigators' entry a continuation of the firefighters valid entry.

The Court also found that the Cliffords' privacy interest in their residence, particularly because they had taken steps to secure it, was greater than the owner's interest in the furniture store in *Tyler.* The Court held that "[at] least where a homeowner has made a reasonable effort to secure his fire damaged home ... we hold that a subsequent postfire search must be conducted pursuant to a warrant, consent, or the identification of some new exigency. So long as the primary purpose is to ascertain the cause of the fire, an administrative warrant will suffice." *Clifford,* at 297. Because no warrant was obtained before the investigators entered the Clifford home, all evidence found during their entry was inadmissible as evidence. Id. at 287.

While not stating it explicitly, the Court implied that even though the investigators were suspicious of arson, an administrative warrant would have sufficed for the initial basement search. The Court noted that even if the investigator's initial search of the basement had been pursuant to a duly authorized administrative search warrant, continuation of the search into the living quarters of the residence was not authorized. Id. at 287. However, because the purpose of an administrative search is to determine the origin of the fire and preclude its rekindling, "not [to] give fire officials license to roam freely throughout the fire victim's private residence," once investigators had determined the cause of the fire was the crock pot and timer they would have exhausted their search authority. Id. at 297-98.

Nevertheless, the one fuel can seized from the driveway by investigators was admissible as evidence. Id. at 299. Because this can was seen in plain view when the investigators arrived at the scene and had been in plain view when the firefighters had entered the house to extinguish the blaze, it was admissible whether it had been seized by the firefighters or the investigators. Id.

[1]By David M. Bessho, Georgia State University, College of Law, Atlanta, Georgia.

Prepared for Michael A. McKenzie; Cozen and O'Connor; Suite 200; One Peachtree Street, NE; Atlanta, Georgia 30308.

RECORDING OF FIRE INCIDENT DATA                    906–27

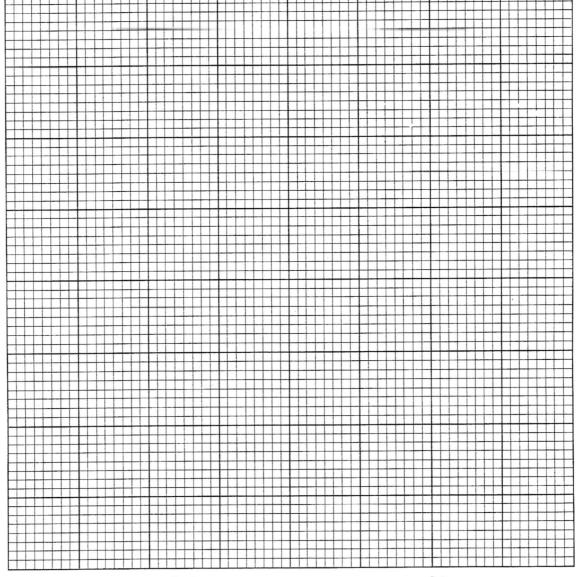

| SKETCH | AGENCY | FILE NUMBER |
|---|---|---|
| FIELD NOTES 906-9 | | |

Scale:                    Sketcher:                    Date:

NOTE: Be sure to show reference north on sketch.

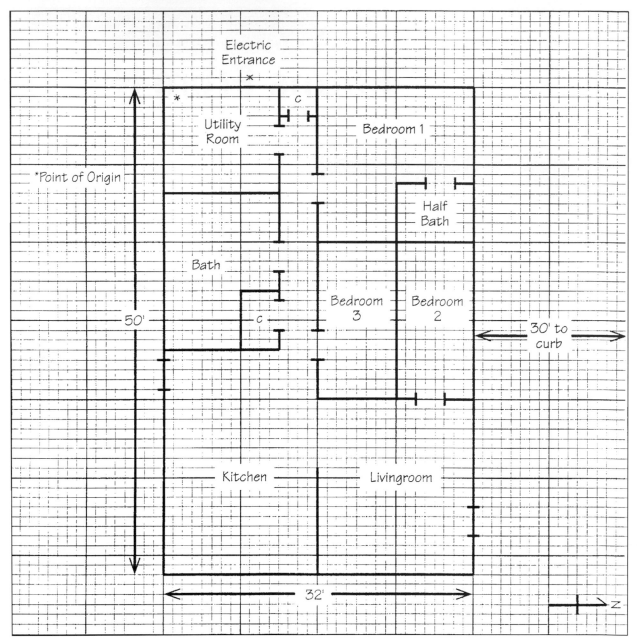

<table>
<tr><td>SKETCH<br>FIELD NOTES 906-9</td><td>AGENCY<br>State Police</td><td>FILE NUMBER<br>1234-93</td></tr>
</table>

Scale: None          Sketcher: Fahey                    Date: 5/27/93

NOTE: Be sure to show reference north on sketch.

| PHOTOGRAPH FIELD NOTES 906-8 | ROLL NO. | AGENCY | FILE NUMBER |
|---|---|---|---|

**\*ONLY ONE ROLL OF FILM PER FORM.**

| NEG. NO. | DESCRIPTION | NEG. NO. | DESCRIPTION |
|---|---|---|---|
| 1 | | 21 | |
| 2 | | 22 | |
| 3 | | 23 | |
| 4 | | 24 | |
| 5 | | 25 | |
| 6 | | 26 | |
| 7 | | 27 | |
| 8 | | 28 | |
| 9 | | 29 | |
| 10 | | 30 | |
| 11 | | 31 | |
| 12 | | 32 | |
| 13 | | 33 | |
| 14 | | 34 | |
| 15 | | 35 | |
| 16 | | 36 | |
| 17 | | 37 | |
| 18 | | 38 | |
| 19 | | 39 | |
| 20 | | 40 | |

**REMARKS**

## PHOTOGRAPH
### FIELD NOTES 906-8

| ROLL NO. | AGENCY | FILE NUMBER |
|---|---|---|
| 1 | State Police | 1234-93 |

*ONLY ONE ROLL OF FILM PER FORM.

| NEG. NO. | DESCRIPTION | NEG. NO. | DESCRIPTION |
|---|---|---|---|
| 1 | Ext – South side | 21 | Doorway to util. rm. from kit. |
| 2 | Ext – East side | 22 | Doorway to util. rm. from util. rm. |
| 3 | Ext – North side | 23 | |
| 4 | Ext – West side | 24 | |
| 5 | Ext – Roof w/hole | 25 | |
| 6 | Ext – S/S near electric ent. | 26 | |
| 7 | Int-- Utility room– S. wall | 27 | |
| 8 | Int– Utility room– E. wall | 28 | |
| 9 | Int– Utility room– N. wall | 29 | |
| 10 | Int– Utility room– W. wall | 30 | |
| 11 | Heater | 31 | |
| 12 | Heater | 32 | |
| 13 | Floor – Utility room | 33 | |
| 14 | Kitchen | 34 | |
| 15 | Livingroom | 35 | |
| 16 | Bedroom #1 | 36 | |
| 17 | Bedroom #2 | 37 | |
| 18 | Bedroom #3 | 38 | |
| 19 | Cleaned floor – Utility room | 39 | |
| 20 | Cleaned floor – Utility room | 40 | |

### REMARKS

To lab for processing 5/27/93

## EVIDENCE
### FIELD NOTES 906-7

| AGENCY | FILE NUMBER |
| --- | --- |

| DESCRIPTION | WHERE FOUND/WHEN | REMOVED TO/BY |
| --- | --- | --- |
| 1. | | |
| 2. | | |
| 3. | | |
| 4. | | |
| 5. | | |
| 6. | | |
| 7. | | |
| 8. | | |
| 9. | | |
| 10. | | |
| 11. | | |
| 12. | | |

### REMARKS

## EVIDENCE
FIELD NOTES 906-7

| AGENCY | FILE NUMBER |
|---|---|
| State Police | 1234-93 |

| # | DESCRIPTION | WHERE FOUND/WHEN | REMOVED TO/BY |
|---|---|---|---|
| 1. | Acme Electric Heater | Utility room floor | Office by Fahey |
| 2. | Debris from near #1 | Utility room floor | Lab by Fahey |
| 3. | Debris from doorway to utility room | | Lab by Fahey |
| 4. | | | |
| 5. | | | |
| 6. | | | |
| 7. | | | |
| 8. | | | |
| 9. | | | |
| 10. | | | |
| 11. | | | |
| 12. | | | |

### REMARKS

Items #2 and #3 taken to lab for analysis to determine presence of flammable liquid 5/27/93.

Item #1 examined in office — found to be on at time of fire 5/27/93.

No flammable liquid present per lab report 5/30/93.

Item #1 destroyed per policy 6/25/93.

# WITNESS STATEMENT
FIELD NOTES 906-6

| AGENCY | FILE NUMBER |
|---|---|
| State Police | 1234-93 |

## IDENTIFICATION

| NAME | ADDRESS | PHONE NO. |
|---|---|---|
| John Ayres | 6478 Smithton Dr. | 686-4913 |

| RACE | SEX | AGE | DATE OF BIRTH | SOC. SECURITY NO. | DRIVER'S LIC. NO. |
|---|---|---|---|---|---|
| W | M | 41 | 4/13/52 | 379-54-7621 | A471-362-479-814 |

| EMPLOYER | ADDRESS | PHONE NO. |
|---|---|---|
| Kentucky Fried Chicken | Clio | 686-0723 |

| RELATIONSHIP TO INCIDENT | CAN BE CONTACTED AT |
|---|---|
| Owner/occupant | Sister – Nancy Singer 684-2239 |

| STATEMENT TAKEN BY | LOCATION, DATE, AND TIME OF STATEMENT |
|---|---|
| Fahey | Scene 5/26/93 11:30 AM |

## STATEMENT

Left for work at 7:45 AM. Thought heater was off, but could have been left on.

Shelf over heater had laundry on it. Also dirty laundry in basket on floor.

No previous fires. He was the last person in the utility room. Heater would start

when temperature dropped to heat room. Cat was killed in fire. All personal

belongings appeared intact.

## WITNESS STATEMENT
### FIELD NOTES 906-6

| AGENCY | FILE NUMBER |
|---|---|
| | |

**IDENTIFICATION**

| NAME | ADDRESS | PHONE NO. |
|---|---|---|
| | | |

| RACE | SEX | AGE | DATE OF BIRTH | SOC. SECURITY NO. | DRIVER'S LIC. NO. |
|---|---|---|---|---|---|
| | | | | | |

| EMPLOYER | ADDRESS | PHONE NO. |
|---|---|---|
| | | |

| RELATIONSHIP TO INCIDENT | CAN BE CONTACTED AT |
|---|---|
| | |

| STATEMENT TAKEN BY | LOCATION, DATE, AND TIME OF STATEMENT |
|---|---|
| | |

**STATEMENT**

# Fire Investigation Incendiary Indicators

Multiple/noncommunicating fires

Unusual fire behavior (fire/flame spread)

Unusual/special burns

Unusual odors

Trailers

Devices

Chemicals and/or flammables

Preparation of premises for fire
- Blocked windows/doors (obscured view)
- Ventilation/fire spread holes in interior
- Absence of furnishings/equipment/stock/pets/etc.

Sabotage of fire fighting equipment
- Hydrants
- Sprinkler systems
- Fire department connections

Nonfunctioning alarm system

Altered/misused electrical equipment or appliances

Evidence of fire to conceal other crime

Access blocked

Similar fires in community/area

Owner(s)/occupant(s) not dressed as expected

Insurance policy immediately available

Vehicles at/leaving scene

Activity (or lack thereof) before fire

Familiar spectators

Building security

Weather conditions

Time of day

General information
- Marital/previous fires/employment/financial

Elimination of accidental causes

# Glossary

## A

**Ampacity**— The amount of electrical current (amperage) that a conductor can safely carry on a continuous basis.

**Ampere** — Basic unit of electrical current that may be abbreviated either by A or I.

**Arc** — A luminous discharge of electricity across a gap. Arcs produce very high temperature.

**Area of Origin** — The room or area where the fire began.

**Artifacts** — The remains of materials involved in the fire that are in some way related to ignition, development, or spread of the fire or explosion.

## B

**Beam**—Structural member subjected to loads perpendicular to its length.

**Bolted Fault** — A condition occurring when two conductors in a circuit come into firm, direct contact with each other. Also called a "dead short."

**Branch Circuit** — The wiring between the point of application (outlets) and the final overcurrent device protecting the circuit.

## C

**Circuit Breaker** — An on/off switch designed to allow a circuit to be opened or closed manually, and to open automatically when a predetermined overcurrent occurs.

## Commodity Flowcharting

**Commodity Flowcharting** — Portrays the movement of a tangible item, such as money or stolen property, through a system.

**Common Conductor** — *See* Grounded Conductor.

**Compression** — Force that tends to push the mass of a material together.

**Conduction** —The point-to-point transmission of heat energy.

**Contamination** — Condition of impurity resulting from mixture or contact with foreign substance. (*Blacks Law Dictionary,* Sixth Edition).

**Convection** — The transfer of heat energy by the movement of heated liquids or gases.

**Crazing** — Formation of patterns of short cracks throughout the pane. It is thought to be the result of heating of one side of a pane while the other side remains cool.

**Current**—The rate of electrical flow in a conductor. Measured in amperes.

## D

**Dead Load** — Load on a structure due to its own weight and other fixed weights.

**Depth of Field** — The range that is in focus in front of and behind the subject of the photograph.

**Due Process** — The requirement that certain procedures that conform with recognized standards of fairness must be followed before a person's life, liberty, or property can be taken away.

## E

**Event Flowcharting**—Chronologically displays the movements of events or occurrences either over time or through a system.

**Evidence** — Information the investigator collects and analyzes.

**Exclusionary Evidence** — Collected to show that a particular device or scenario can be ruled out with relation to the ignition or fire spread scenario.

## F

**Fire Cause** — The sequence of events that allows the source of ignition and the fuel to come together.

**Fire Patterns** — Visible or measurable physical effects that remain after a fire.

**Focal Length** — The focal length of a lens is the distance behind the lens where light from an object is sharply focused when the lens is set to infinity.

**Footing**—The part of the building that rests on the bearing soil and is wider than the foundation wall. Also the base for a column.

**Fuse** — A single-acting protective device designed to open a circuit on a predetermined overcurrent.

## G

**Girder**—Large, horizontal structural member used to support the ends of joists and beams.

**Grounded Conductor**—That conductor in a branch circuit which carries the return current but which is not energized. Often called the common or neutral conductor. The covering will be white or natural gray in color. It will be connected to the service neutral.

**Ground Fault** — A current that flows to ground outside of the normal current path, such as through a ground conductor, metal pipes, or a person.

**Ground-Fault Circuit Interrupter (GFCI)** — A device designed to protect people from electrical shock by opening (de-energizing) a circuit when a current to ground exceeds a predetermined value but that is still at less than life-threatening levels.

**Grounding Conductor** — That conductor in a branch circuit that connects exposed metal parts of appliances to the ground system of the service to minimize the chance of electric shock. This conductor carries no current unless a fault has occurred. The conductor will be bare or have a green or green with yellow-stripe covering.

## H

**Heat** — Energy transferred between two bodies of differing temperature, such as the sun and the earth.

**Heat of Ignition** — The heat energy that brings about ignition. Heat energy comes from various forms and usually from a specific object or source. Therefore, the heat of ignition is divided into two parts: (a) equipment involved in ignition and (b) form of heat of ignition.

**Hot Conductor** — *See* Ungrounded Conductor.

## I

**Incendiary**—A fire deliberately set under circumstances in which the responsible party knows it should not be ignited.

**Inference Development** — The development of a meaningful hypothesis, conclusion, prediction, or estimate based on the data available and the knowledge, experience, and expertise of the investigation.

**Interrogation**—A formal line of questioning of an individual who is suspected of committing a crime or who may be reluctant to provide answers to the investigator's questions.

**Interrupt Rating** — The highest current at rated voltage that a device is intended to interrupt under standard test conditions.

**Interview** — The questioning of an individual for the purpose of obtaining information related to the investigation.

## J

**Joist**—Framing member that directly supports the floor.

## K

**Kinetic Energy** — The energy possessed by a moving object.

## L

**Layering** — The deposition of fire debris in identifiable layers that is above or below a floor assembly, ceiling materials, or roof assembly.

**Light** — Visible radiation produced at the atomic level, such as a flame produced during the combustion reaction.

**Link Analysis** — Method of computing, organizing, and utilizing data relating to an investigation. It allows the analysis and presentation of complex data in a clear and concise manner.

**Live Load** — Furniture, people, or other movable loads not included as a permanent part of the structure.

**Load** — The sum of the wattages of the various devices being served by a circuit.

## M

**Material First Ignited** — The fuel that is first set on fire by the heat of ignition. To be meaningful, both a type of material and a form of material should be identified.

**Matter** — Anything that occupies space and has mass.

## N

**Neutral Conductor** — *See* Grounded Conductor.

## O

**Ohm** — The basic unit of electrical resistance symbolized either by W or R. One ohm is the resistance between two points in a conductor when one volt produces one ampere of current.

**Ohm's Law** — The mathematical relationship between a circuit's voltage (V), current (I), and resistance (R): $V = IR$.

**Overcurrent** — Any current that is in excess of the rated current of equipment or the ampacity of a conductor and may be caused by an overload, short circuit, or ground fault.

**Overload** — Operation of equipment or a conductor in excess of its rated ampacity. If continued for a sufficient length of time, overheating to the point of damage may occur.

## P

**Panelboard** — Single or multiple panels that contain conductive bus bars and automatic overcurrent protection devices such as circuit breakers or fuses. These panels may also contain manually operated switches.

**Parapet** — Low wall at the edge of a roof.

**Partition** — Interior wall that separates a space into rooms.

**Point of Origin** — The exact physical location where the heat source and fuel come in contact with each other and a fire begins.

**Potential Energy** — Energy possessed by an object that can be released in the future.

**Power** — Amount of energy delivered over a given period of time.

## R

**Radiation** — The transmission of energy as an electromagnetic wave (such as light waves, radio waves, or X rays) without an intervening medium.

**Resistance** — The opposition to the flow of an electric current in a conductor or component. Measured in ohms (W).

## S

**Shear Stress** — Stress resulting when two forces act on a body in opposite directions in parallel adjacent planes.

**Sheathing** — Covering applied to the framing of a building to which siding is applied.

**Short Circuit** — An abnormal path of current in a circuit that normally leads to an overcurrent condition.

**Slab** — (1) Heavy steel plate used under a steel column. (2) The reinforced concrete floor itself.

**Spalling** — Expansion of moisture within concrete due to exposure to the heat of a fire. The expansion causes sections of the concrete surface to violently disintegrate.

**Spoliation** — The intentional destruction of evidence, and when it is established, fact finder may draw inference that evidence destroyed was unfavorable to party responsible for spoliation. (*Blacks Law Dictionary,* Sixth Edition)

## T

**Tensile** — Force of pulling apart or stretching.

**Trailer** — Often used in intentionally set fires to connect remote fuel packages (combustible materials, pools of ignitable liquid, etc.) with each other.

**Trip Curve** — The relationship between current and time that determines when circuit protective devices operate. (This data is provided by the manufacturer.)

## U

**Ungrounded Conductor** — That conductor in a branch circuit which is energized and supplies current to load devices. The covering will be some color other than white, gray, or green. Often called the hot conductor.

## V

**V-pattern** — One of the patterns often found during an interior examination. A cone-shaped plume of hot gas that forms above the flames.

**Volt** — The basic unit of electrical voltage and may be abbreviated either V or E. It is the difference in potential (electromotive force) needed to create a current of one ampere through the resistance of one ohm.

**Voltage** — The electrical force that causes a charge (electrons) to move through a conductor. Sometimes called the electromotive force (EMF). Measured in volts (V)

## W

**Watt** (W) — The basic unit of power. In simple electrical systems, power is equal to voltage times current. One watt is equal to a current of one ampere under the potential of one volt.

# Index

service wires (electrical) in buildings, 75–76
shafts (ductwork, piping, wiring), 60
shipping papers for cargo, 92
shock hazards, 9
short circuits, 183
simultaneous ignition, 32
sketches of fire scene, 108–110, 215–216
smoke, 36–37
smoking materials as ignition source, 185–186
solder as ignition source, 181
solid fuel equipment as ignition source, 182
solid fuels, 24–25, 179
sources of ignition, 179–188, 203
spalling (concrete), 48, 195
sparks as ignition source, 180–181, 185–186
spoliation, 103, 104
spontaneous heating, 27–28, 186
spreading of fire, 30–34
static electricity, 187–188
steel, 43, 48
steel-reinforcing bars, 47
step-down transformers, 73–74
stone building materials, 47, 53
summary offenses, 205–206
sunlight as ignition source, 186
Supreme Court cases, 211–213
suspects, interviewing, 147–149
suspicious fires, 190
system components, examining, 159
systematic approach to fire examination, 97

**T**
tape (adhesive) evidence, 130–131
target fuels, 34
telecommunication cables, 60
tempered glass, 49
tension rings, 51
terrorism as motive for arson, 197
testimony, 205, 206–207
thermal layering of gases, 34–35
time line graphs, 193–194
tire prints, preserving as evidence, 136–137
tool marks, preserving as evidence, 136–137
tools of fire investigators, 10–12
trailers, 158, 169, 194
transformers, step-down, 73–74
transportation and conveyance systems, 60–61
trials, felony, 206
trip curves (circuit breakers), 81
trusses (roof), 51
"Tyler" case. *See* Michigan *vs.* Tyler case
Type I (fire-resistive) construction, 42
Type II (noncombustible) construction, 42–43
Type III (ordinary) construction, 43

Type IV (heavy timber) construction, 43–44
Type V (wood-frame) construction, 44–45

**U**
U-shaped fire patterns, 166
undetermined fire causes, 190
Uniform Hazardous Waste Manifest, 91–92
unprotected Type II construction, 42
upper flammable limit (UFL), 26
utilities (electric, gas, water) in buildings, 9

**V**
V-patterns, 157, 165–166
vandalism as motive for arson, 195
vanity as motive for arson, 196–197
vapor density of gases, 21
vaporization of liquid fuel, 27
vector diagrams, 171–172
vehicle fire reports, 202
vehicles, collecting electrical system evidence, 134
ventilation-controlled fires, 31
ventilation-generated fire patterns, 167–168
ventilation in self-ignition process, 187
video photography, 121
volatility of liquids, 25

**W**
walk-through process during interior examination, 162
wall and ceiling coverings, building code classifications, 53
water damage, 155
wicking effect in fire patterns, 169
wildfires, information in reports, 202
wired glass as building material, 49
wiring
    electrical resistance, 70
    parallel and series circuits, 66–70
    protecting from physical damage, 60
    required sizes in branch circuits, 71–72
    splices and connections, 78–79
witnesses
    expert, 207
    of fact, 207
    interviewing, 147–149
    separating, 141
    statements from, 144, 203, 221–222
wood
    in building construction, 47
    collecting as evidence, 132
    ignition times, 181
wood-frame (Type V) construction, 44–45
wood shingles, igniting, 52

*Indexed by Karen Lane.*

**Fire Investigator**
**First Edition**

**COMMENT SHEET**

DATE _____ NAME _____

ADDRESS _____

ORGANIZATION REPRESENTED _____

CHAPTER TITLE _____ NUMBER _____

SECTION/PARAGRAPH/FIGURE _____ PAGE _____

1.  Proposal (include proposed wording or identification of wording to be deleted),
       OR PROPOSED FIGURE:

2.  Statement of Problem and Substantiation for Proposal:

RETURN TO:  IFSTA Editor          SIGNATURE _____
            Fire Protection Publications
            Oklahoma State University
            930 N. Willis
            Stillwater, OK 74078-8045

Use this sheet to make any suggestions, recommendations, or comments. We need your input to make the manuals as up to date as possible. Your help is appreciated. Use additional pages if necessary.

**COMMENT SHEET**

DATE _____ NAME _____

ADDRESS _____

ORGANIZATION REPRESENTED _____

CHAPTER TITLE _____ NUMBER _____

SECTION/PARAGRAPH/FIGURE _____ PAGE _____

1. Proposal (include proposed wording or identification of wording to be deleted),
     OR PROPOSED FIGURE:

2. Statement of Problem and Substantiation for Proposal:

RETURN TO:  IFSTA Editor             SIGNATURE _____
              Fire Protection Publications
              Oklahoma State University
              930 N. Willis
              Stillwater, OK 74078-8045

Use this sheet to make any suggestions, recommendations, or comments. We need your input to make the manuals as up to date as possible. Your help is appreciated. Use additional pages if necessary.

# Your Training Connection.....

**ifsta**

The International Fire Service Training Association

We have a free catalog describing hundreds of fire and emergency service training materials available from a convenient single source: the International Fire Service Training Association (IFSTA).

Choose from products including IFSTA manuals, IFSTA study guides, IFSTA curriculum packages, Fire Protection Publications manuals, books from other publishers, software, videos, and NFPA standards.

Contact us by phone, fax, U.S. mail, e-mail, internet web page, or personal visit.

*Phone*
1-800-654-4055

*Fax*
405-744-8204

*U.S. mail*
IFSTA, Fire Protection Publications
Oklahoma State University
930 North Willis
Stillwater, OK 74078-8045

*E-mail*
editors@ifstafpp.okstate.edu

*Internet web page*
www.ifsta.org

*Personal visit*
Call if you need directions!